Un
Po
and Standby
Power Systems

I0034035

Uninterruptible Power Supplies and Standby Power Systems

Alexander C. King

William Knight

McGraw-Hill

New York   Chicago   San Francisco   Lisbon   London
Madrid   Mexico City   Milan   New Delhi   San Juan
Seoul   Singapore   Sydney   Toronto

The McGraw·Hill Companies

Cataloging-in-Publication Data is on file with the Library of Congress

Copyright © 2003 by The McGraw-Hill Companies, Inc. All rights reserved. Printed in the United States of America. Except as permitted under the United States Copyright Act of 1976, no part of this publication may be reproduced or distributed in any form or by any means, or stored in a data base or retrieval system, without the prior written permission of the publisher.

1 2 3 4 5 6 7 8 9 0   DOC/DOC   0 8 7 6 5 4 3 2 1

ISBN 0-07-139595-4

The sponsoring editor for this book was Steve Chapman, the editing supervisor was Steven Melvin, and the production supervisor was Sherri Souffrance. It was set in Century Schoolbook following the MHT design by Joanne Morbit and Paul Scozzari of McGraw-Hill Professional's Hightstown, N.J., composition unit.

RR Donnelley was printer and binder.

Chap. 1: Extracts from both ISO and British Standards are reproduced with the permission of BSI under license number 2001 SK/041 0. Complete standards can be obtained from 1351 Customer Services, 389 Chiswick 111gb Road, London, W4 4AL. (Tel. +44 (0) 020 8996 9001). U.K. crown copyright is reproduced with the permission of the Controller of Her Majesty's Stationery Office.

McGraw-Hill books are available at special quantity discounts to use as premiums and sales promotions, or for use in corporate training programs. For more information, please write to the Director of Special Sales, McGraw-Hill, 2 Penn Plaza, New York, NY 10121-2298. Or contact your local bookstore.

This book is printed on recycled, acid-free paper containing a minimum of 50% recycled, de-inked fiber.

Information contained in this work has been obtained by The McGraw-Hill Companies, Inc. ("McGraw-Hill") from sources believed to be reliable. However, neither McGraw-Hill nor its authors guarantee the accuracy or completeness of any information published herein and neither McGraw-Hill nor its authors shall be responsible for any errors, omissions, or damages arising out of use of this information. This work is published with the understanding that McGraw-Hill and its authors are supplying information but are not attempting to render engineering or other professional services. If such services are required, the assistance of an appropriate professional should be sought.

Contents

Preface

The authors of this book have been associated with uninterruptible and standby power supplies for many years. The origin of the book could be said to go back to the 1980s when the authors became involved in various training lectures for engineers employed within the Property Services Agency (no longer in existence) of the U.K. government.

There followed a demand for commercially based lectures aimed at midcareer engineers from which it became apparent that a published book would be helpful to those persons who became involved in high-quality power systems.

So it was decided to publish this book, the intention being to present to interested persons a picture of the wide variations in design now available for uninterruptible power supplies and, to a lesser extent, for standby generating plants. It would provide advice on the best way forward at the planning stage and on the pitfalls to be avoided.

There is a great variety of basic types of uninterruptible power supply systems and the details of design are forever changing. Equipment installed a few years ago will differ from what is installed today, and engineers working in the field will encounter a great variety of equipment. For this reason the text explains the historical development of the systems, the idea being to assist readers who may encounter older equipment which has been operating for several years. The history and origin of circuits also enables the reader to see the direction in which future developments are likely to occur.

The generation of standby power does not lend itself to innovation to the same extent, most standby power generating sets are, and always have been, driven by diesel engines. The diesel engine has of course changed over the years and modern engines are technically superior to their predecessors.

The name used for uninterruptible power supplies is a misnomer. There really is no such thing! Reliability follows from a detailed examination of the complete system, its environment, and its maintenance.

One of the main reasons for failure is human error. However, a remarkably high standard of reliability can be achieved for both uninterruptible and standby power supplies and the authors hope that the contents of this book will assist readers to achieve that objective.

Looking back, the UPS market had its early beginnings nearly 60 years ago, and naturally there have been tremendous advances since then in reliability, efficiency, dimensions, and cost. But developments continue in a thriving and ever-expanding market, and under such conditions one should expect a healthy development program. Indeed such is the case, we can see small gas turbines coming into the marketplace offering ease of installation and low NO_X levels. Battery development shows a future where designs will reflect the increasing diversity in UPS design and application. Batteries will, in the future, themselves possibly use other chemical formulae (lithium-ion), and on the horizon one can see fuel cells being developed for use with UPS systems.

We believe the development of static UPS will continue as more diverse and advanced solid state switching components become available. The rotary system, the earliest type of UPS, was at one time completely eclipsed by static systems but its more sophisticated successors have now regained a place in the market and will continue to increase their effectiveness. Expect to see developments at the lower end of power ratings.

Bill Knight has written the text on uninterruptible power supplies, Alec King has written the text on standby power supplies and the chapter on harmonics. This will explain the different styles of the two parts of the book.

Both authors take this opportunity to thank the many colleagues, too numerous to name, who have provided information, given their time, or otherwise encouraged us to complete the book.

A. C. King
W. R. Knight

Standby Power Generating Sets

Introduction

This chapter briefly discusses where and why the need for standby generation arises and describes the systems that are included in a normal
standby power generating set. Most standby generating sets are diesel
engine driven and this book concentrates on such sets. A small number of
sets above about 500 kW may be driven by gas turbines and the section
titled "The Power Unit" includes an introduction to gas turbines. They are
mentioned elsewhere in the text but their use and characteristics are not
described in such detail as are the use and characteristics of diesel engines.

The Need for Standby Generation

The need for standby generation arises if the consequences of a failure or
disruption of the normal supply are not acceptable. The types of installation in which the need arises seem to be limitless. There are basically four
reasons for installing standby generation: safety, security, financial loss,
and data loss.

Safety Where there is a risk to life or health such as in air traffic control, aviation ground lighting, medical equipment in hospitals, nuclear
installations, oil refineries

Security against vandalism, espionage, or attack Area lighting, communication systems, military installations, etc.

Data loss Situations in which the loss of data may be catastrophic
and irretrievable such as data processing and long-term laboratory
type of testing or experiment

Financial loss Critical industrial processes, large financial institu-
tions, etc.

Standby generation is often installed to provide a long-term back-up to an uninterruptible power supply which will have been installed for one of the reasons mentioned above.

The Generating Set and Its Supporting Systems

The major components of a generating set are the power unit and the generator; these are considered in the sections titled "The Power Unit" and "Alternating Current Generators" which follow. The remaining sections of this chapter are devoted to the many supporting components and systems such as speed governors, voltage regulators, cooling and fuel systems, ventilating and exhaust systems. Many of these components will include a control system which may operate independently or may be linked to other control systems, thus the cooling system will initiate an over-temperature shut down procedure, the mains monitoring system will initiate a starting procedure, and the loss of one set in a multiset installation may initiate load shedding.

The International Standard for diesel driven generating sets is ISO 8528—Reciprocating internal combustion engine driven alternating current generating sets. This is a comprehensive document containing a wealth of information and well worth studying for anyone wishing to acquire detailed information about generating sets.

There is no equivalent standard for gas turbine–driven generating sets.

The Power Rating Classification of Diesel Engine–Driven Generating Sets

Rating classes applicable to diesel engine–driven generator sets are described in ISO 8528 and are discussed in the following paragraphs. None of the ratings include any overload capacity.

Continuous Power (COP)

Continuous power (Fig. 1.1) is the power which the set can deliver continuously for an unlimited number of hours per year between the stated maintenance intervals.

Prime Power (PRP)

This rating is applicable to sets supplying a variable power sequence. The sequence may be run for an unlimited number of hours per year between the stated maintenance intervals. Prime power is the maximum power generated during the sequence and the average power over

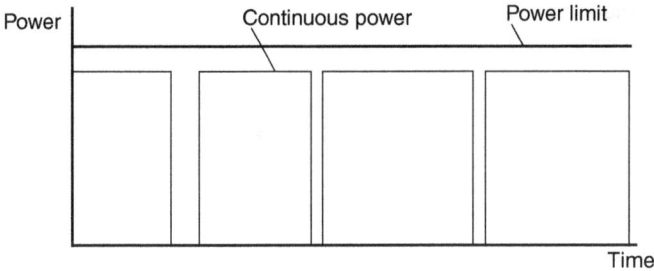

Figure 1.1   Illustration of continuous power.

any 24-hour period is not to exceed a stated percentage of the prime power. In calculating the average power, powers of less than 30 percent shall be taken as 30 percent and any time at standstill shall not be counted. The example of generating set sizing which appears in Chap. 3 uses prime power rating.

As the 24-hour average power of a PRP-rated set is increased, it becomes closer to a COP rating; if the average power is equal to the prime power the set would in effect be rated for continuous power.

This rating is suitable for standby supply generating purposes. The prime power is available for peak loads which occur after start-up such as motor starting and UPS battery charging. and after these loads have reduced, the steady state load remains (Fig. 1.2). The 24-hour average of all these loads is calculated and must not exceed the agreed percentage of prime power that may be used. It should be noted that during any 24-hour period there may be several supply failures, each of which will increase the average power loading if peak loads occur after each start-up.

In the illustration the average power over the 24-hour period may be calculated from the formula:

$$\text{Average power} = \frac{P_1 t_1 + P_2 t_2 + P_3 t_3 + P_4 t_4 + P_5 t_5}{t_1 + t_2 + t_3 + t_4 + t_5} \tag{1.1}$$

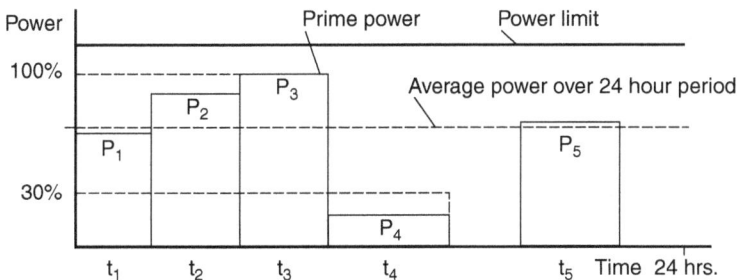

Figure 1.2   Illustration of prime power rating.

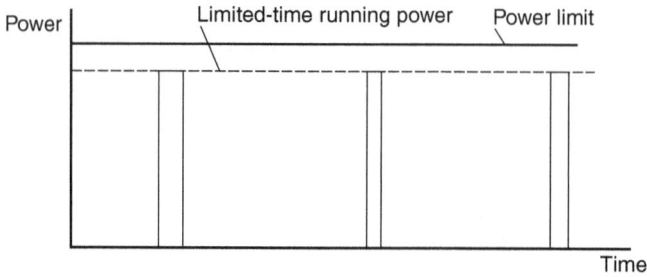

Figure 1.3   Illustration of limited-time running power.

Limited-Time Running Power (LTP)

Limited-time running power (Fig. 1.3) is the maximum power which a generating set is capable of delivering for up to 500 hours per year, of which a maximum of 300 hours is continuous running, between the stated maintenance intervals. It is expected that the periods of running will be long enough for the engine to reach thermally stable conditions. This may be suitable as a low-cost option for standby supply generating purposes. It differs from the PRP rating in that it allows the engine and the generator to run at full capacity for the permissible running time. The engine wear rate will be greater, the generator will run hotter and the insulation will deteriorate at a faster rate. The set will therefore have a shorter life expectancy than a more conservatively rated set.

Power Limit

In addition to the three power ratings described above, ISO 8528 recognizes a power limit, determined by the fuel rack stop on the engine fuel-injection system, which is greater than the power required to satisfy the ratings. When the engine is supplying its maximum power to the generator, this surplus power is available for governing purposes and is necessary if speed is to be maintained within correct limits.

The Power Unit

Diesel Engines

The development of diesel engines has progressed steadily over many years due to improved techniques and knowledge of machining, lubrication, metallurgy, combustion, and noise and vibration control.

Engines are available from a few kilowatts upwards at speeds of 750, 1000, and 1500 rpm for 50-Hz supplies and 900, 1200, and 1800 rpm for 60-Hz supplies. Reliability and cost reduce with increasing speed. Standby sets up to about 1.5 MVA may run at 1500 rpm provided that a long-term base load type of operation is not envisaged. Some installations may be required to continue running for long periods after normal supplies have failed (e.g., some military installations) and in such cases 1000-rpm engines should be considered above approximately 1 MW. Most diesel engines in use are of four-stroke type, but two-stroke engines may occasionally be encountered. Diesel engines are designed to run on Class A fuel to BS2869 which has a calorific value of about 42.7 MJ/kg; before running on any other fuel, advice should be obtained from the engine manufacturer. The efficiency of a modern turbocharged engine may be about 40 percent but this does not take into account any auxiliary drives or the generator losses; the overall efficiency of the generating set will be less.

The useful energy produced by the engine passes through the coupling to the generator but, depending on the arrangement of the set, it is not always possible to use all the energy to supply the intended load. Sets up to a few hundred kVA are usually self-contained but larger sets may require power for auxiliary items such as cooling and ventilation fans. The requirement will be small, only a few percent of the generator rating. There may also be incidental extras such as engine room lighting and small power and fuel pumps.

A naturally aspirated diesel engine is capable of accepting full load in a single step but, in order to reduce the size and cost, many modern engines are turbocharged. This reduces the step load capability and a modern turbocharged engine with a high brake mean effective pressure (see below) will probably accept only 60 or 70 percent of its rated load in one step. It follows that for most installations a load switching sequence has to be followed after the starting procedure.

With a naturally aspirated engine the quantity of combustion air within the combustion space is constant and there is always sufficient oxygen for combustion of the maximum amount of fuel. With a turbocharged engine the quantity of surplus combustion air available at any one time is limited by the turbocharger. A step load change may cause a sudden increase in the amount of fuel injected but there will be inadequate combustion air until the turbocharger has had time to accelerate.

Generating set manufacturers recognize four categories of load acceptance and categories 1, 2, and 3 are typical of the sets used for standby generation:

Category 1	100 percent
Category 2	80 percent
Category 3	60 percent
Category 4	25 percent

The load acceptance is closely related to the brake mean effective pressure (BMEP) of the engine. The BMEP is derived from the mechanical power developed by the engine, its speed, number of cylinders, and the swept volume per cylinder, from the relationship:

$$\text{BMEP} = \frac{\text{Engine brake power}}{\text{Swept volume of each cylinder} \times \text{Firing strokes per second}}$$

$$(1.2)$$

It follows that the BMEP is related to the compression ratio and the degree of turbocharging. ISO 8528-5 includes guide values for step loading as a function of BMEP.

Subject to the step load limitations, a diesel engine will be ready to accept load within 10 to 15 seconds of receiving its start signal. To ensure that the set is ready for immediate starting, it is usual to include a jacket water heater or heaters which ensure that the bearing surfaces and cylinder bores do not unduly cool the oil during starting and initial running. In winter the heat introduced into the engine from these heaters reduces the heat required to maintain the engine room temperature.

It is also usual to incorporate a lubricating oil priming system which ensures that the engine mating surfaces are wet before the crankshaft is turned for starting. A continual priming cycle is usually adopted, say a few minutes every hour, to maintain the surfaces in a condition suitable for cranking. As soon as the engine is up to speed, the main oil pump takes over the duty and the priming system is shut down. For large engines the alternative of continuous priming, as distinct from a continual cycle, can be used but there is a danger between test runs of oil draining down a valve stem and collecting above a piston. At the next start this could lead to hydraulic blocking and engine damage.

The reciprocating masses of an engine lead to vibration of the main frame which must be isolated from its mountings. Assuming that the engine and generator are solidly bolted together to form a single mass, the usual arrangement is for the generating set to be supported by vibration dampers fixed to a base frame which rests on the engine room floor. If the engine and generator are separately mounted and connected with a flexible coupling, a base frame will be required to support the generating set and this will be supported by vibration dampers either fixed to a sub-base as described above, or resting directly on prepared mounting pads at floor level. Some older installations may include engines solidly bolted to massive concrete blocks independently supported and isolated from the floor by cushioning material such as cork, but any such installations will be approaching obsolescence.

Standard Reference Conditions for Diesel Engines

The standard conditions for diesel engines are specified in ISO 8528-1 and ISO 3046-1 as:

Total barometric pressure	100 kPa (1 bar)
Air temperature	25°C
Charge air coolant temperature	25°C
Relative humidity	30 percent

The barometric pressure of 100 kPa is equivalent to an altitude of 150 meters above sea level. The altitude at which an engine is working has an important effect on the engine's performance. At high altitudes a smaller mass of aspiration air is drawn into each cylinder, less fuel can be burnt and less power is produced.

The ambient air temperature of 25°C imposes some limitation of output in temperate and warm climates. It is usual to allow a 10°C rise in the engine room ventilating air and if the outside ambient exceeds 15°C the temperature within the engine room is likely to exceed 25°C and some degree of derating will be necessary. The engine manufacturer should be made aware of the maximum operating temperature expected and of its duration.

Gas Turbines

Gas turbines are currently available from about 500 kW upwards; at the time of going to press units as small as 50 kW were being considered for combined heat and power purposes so there may be a future downward trend in the size of gas turbines for standby power applications. The turbine shaft will run at tens of thousands of rpm and a gear box is interposed between the turbine and the generator, which will run at 1500 or 3000 rpm for 50-Hz supplies.

There are two types of gas turbine, the single-shaft machine and the two-shaft machine. The single-shaft machine has the turbine and the compressor on a single shaft, and is used for standby generation applications. The turbine and the compressor run at constant speed and the mass flow of air through the machine is constant. There is always adequate combustion air and any increase of fuel leads to an immediate increase of power; full load can be accepted in a single step.

The two-shaft machine is of higher efficiency but is not used often for standby generation purposes because it has a poor step load acceptance. The turbine and the compressor are on separate shafts, the compressor being driven by the turbine exhaust gas flow. The speed of the compressor and mass flow of air through it is variable; any increase of load requires an increase of fuel, which in turn requires an increase of combustion air, but this is not available until the compressor has attained an increased speed. This is similar to the situation encountered with turbocharged diesel engines as previously described.

The efficiency depends on the turbine parameters, for a single-shaft machine at full load it may approach 25 percent which leads to a fuel usage of 0.45 liters per kWh or 1 liter per 2.2 kWh. The efficiency will drop off rapidly at lower loads due to the constant air flow characteristic. The constant air flow results in a constant compressor loading and the surplus combustion air at low loads leads to a lower operating temperature. At 50-percent loading the efficiency may drop to 15 percent, the manufacturer should be consulted for a realistic figure. For standby generation purposes the efficiency is of less importance than for continuous running applications although a low efficiency leads to a large fuel storage requirement. It is possible, however, to design a gas turbine to run on almost any fuel, it may therefore be possible to make use of boiler fuel or whatever is available on site. It is of course possible to run on gas but it is not easily stored and the supply may not be reliable.

Gas turbines do not have any reciprocating masses and have minimal vibration due to dynamic unbalance. Cooling is achieved mainly by the combustion air flow, a simpler arrangement than the jacket water cooling for diesel engines. The main bearings are cooled by the lubricating oil which passes through an oil/air or oil/water heat exchanger.

A gas turbine will be very much smaller than a diesel engine of the same power rating but it is not easy to quantify the difference between installations because the gas turbine will require more space for its air inlet and exhaust systems and will probably require additional acoustic treatment. Any restriction of the flow of the inlet air or of the exhaust gases has a significant effect on the performance of the gas turbine. There will be an air inlet filter and air inlet and exhaust acoustic attenuators. In order to achieve low pressure drops these will require large cross-sectional areas, leading to large inlet and exhaust ductwork. The ductwork is normally designed as part of the gas turbine package and the air inlet pressure drop is likely to be of the order of 1 kPa (100-mm water gauge). The exhaust pressure drop may be a little less at say 0.75 kPa (75-mm water gauge).

Gas turbines require specialized maintenance and the necessary skills are not so widely available as they are for diesel engines.

Standards Relating to Gas Turbines

There is no international standard applicable to gas turbine–driven generating sets, nothing equivalent to ISO 8528. There are two standards of particular interest, ISO 3977—Gas Turbine Procurement and ISO

2314—Specification for Gas Turbine Acceptance Tests, which are briefly discussed below. There are other international or British Standards relating specifically to gas turbines and noise measurement, exhaust gas emissions, vibration, fuels, a glossary of terms, and graphic symbols.

ISO 3977—Gas Turbine Procurement lists a variety of operational modes for annual running hours and for the number of starts per annum. There are four classes for annual running hours: Class A allows for up to 500 hours per annum and would suit most standby applications, other classes allow for up to 2000, 6000, and 8760 hours per annum. There are five ranges for the number of starts per annum: up to 25, up to 100, up to 500, above 500, and continuous operation. Range III allows for up to 100 starts per annum and would be suitable for most standby applications; Range IV allows for 25 starts per annum which would restrict the number of test runs. The standard includes other sections on fuel, control and protective devices, vibration, sound, and pollution. There is a section listing the technical information to be supplied by the purchaser with the enquiry, and another on the technical information to be supplied by the manufacturer when tendering.

ISO 2314—Specification for Gas Turbine Acceptance Tests describes in some detail the procedures to be followed when undertaking acceptance tests which may be on a complete generating set including the generator.

Standard Reference Conditions for Gas Turbines

The standard conditions for gas turbines are specified in ISO 2314 and ISO 3977 as:

1. Intake air at compressor flange
   a. A total pressure of 101.3 kPa
   b. A total temperature of 15°C
   c. A relative humidity of 60 percent
2. Exhaust at turbine exhaust flange
   a. A static pressure of 101.3 kPa

The barometric pressure of 101.3 kPa is equivalent to operating at sea level. The temperature of 15°C imposes some limitation in temperate and warm climates. The manufacturer should be aware of the altitude at which the turbine will operate and of the maximum expected ambient temperature.

Alternating Current Generators

Power passes from the engine through the coupling to the generator which will take one of two forms:

- A single-bearing generator with the frame spigot mounted directly to the engine crankcase and the driven end of the generator shaft supported by the engine crankshaft via a coupling

- For larger sizes, a two-bearing generator may be used. The engine and generator are solidly mounted on a base frame and the generator is driven through a flexible coupling.

Diesel engine–driven generators will run at engine speed (1000 or 1500 rpm for 50 Hz) and will have salient pole rotors. Gas turbine–driven generators will usually run at 3000 rpm (for 50 Hz) and will probably have cylindrical rotors.

The distribution voltages in common use within the United Kingdom are 400 and 11,000 volts and generators will usually use one of these voltages. It is not generally economical to manufacture a high voltage machine for ratings below about 1 MVA, so below this size generation may be expected to be at low voltage. The economics depend on the material content of the stator windings, a low voltage machine with a high rating would include an excessive amount of copper and a small amount of insulation, whereas for a high voltage machine with a low rating the reverse situation would apply.

However, a rating of 1 MVA at 400 volts results in a line current of 1443 amperes, and cables for such a current are quite large and may be unmanageable. If cabling is likely to be troublesome due to heavy currents, consideration can be given to generating at low voltage and adding a generator transformer. Generator transformers are discussed later in this section.

Excitation Systems

Modern generators use brushless excitation systems, but there remain in use many machines provided with dc exciters having commutators and brushgear. The advent of semiconductor rectifiers made it possible to replace the dc exciter with a much simpler ac exciter and a rectifier mounted on the generator shaft. This arrangement dispenses with the brushgear and its attendant maintenance problems and is achieved at lower cost.

The ac exciters now fitted require a power supply to energize the stator field and there are two methods in use:

- A permanent magnet pilot exciter provides the field supply for the main exciter as shown in Fig. 1.4.

- The exciter field takes a supply from the generator output as shown in Fig. 1.5. Note that for high voltage machines this involves the addition of a step down transformer; it is therefore not recommended for high voltage machines.

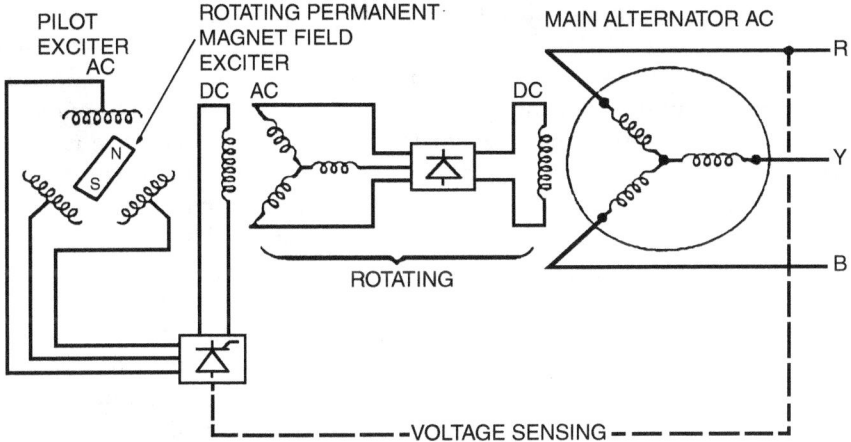

Figure 1.4 Excitation system using a pilot exciter.

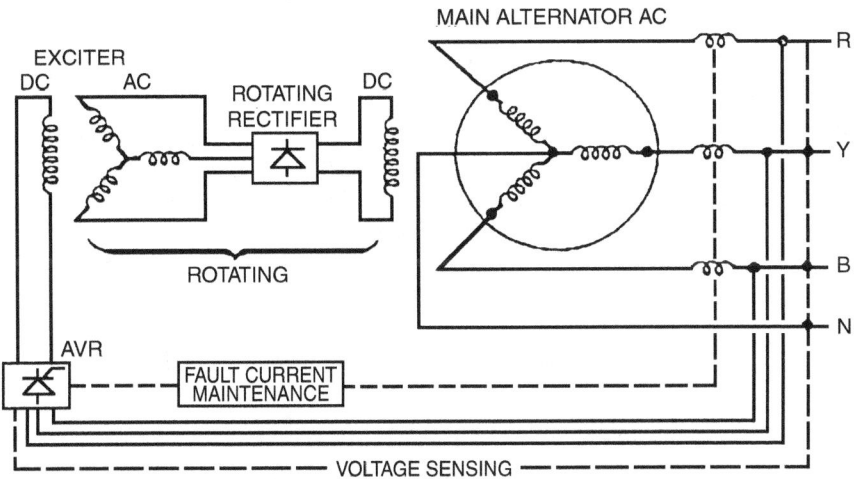

Figure 1.5 Excitation system taking a supply from the generator output.

The permanent magnet pilot exciter has the advantage that it provides a supply to the main exciter field that is independent of the generator output. If the generator output suffers a short circuit the field remains excited and the generator is able to deliver its short circuit current until protective equipment operates to clear the fault.

Where there is no pilot exciter, power for the exciter field is taken from the generator output terminals. With this arrangement, unless precautions are taken, a fault on the distribution system will lead to a loss of voltage on one or more phases, which in turn can lead to a loss of excitation power and to total collapse of the generator output. In such circumstances the fault may not be cleared by the protection equipment and the standby supply may be out of use until the fault is removed manually from the system. The probability of this happening is affected by whether one, two, or three generator phases are used to supply the field power, and by which phases are affected by the fault.

To avoid a collapse of the generator output, a current transformer may be added to each phase of the generator output. The current transformer secondary windings are connected to the excitation system and maintain the field power until the fault is cleared by the protective equipment. There is an alternative system, less widely encountered, which is applicable only to salient pole machines and uses a separate stator winding specially to back up the field supply during faults. This winding is distributed in the stator to take advantage of the third harmonic voltages which appear when heavy currents flow from a salient pole generator. The third harmonic voltages provide the field power during faults in the same way as the previously described fault current maintenance current transformers.

Pole Face Damper Windings

Salient pole generators should be fitted with pole face damper windings having interconnections between the poles. These windings improve the waveform by attenuating harmonic fluxes in the machine air gap; where machines are to be run in parallel they are essential to prevent phase swinging, the phenomena in which the machine rotors oscillate in angular rotation (and position) either side of the desired position. In this condition the damper windings are in motion relative to the stator flux, voltages are induced and the resultant current flow has a damping effect.

Generator Transformers

A generator transformer is used where the voltage of the generator differs from that of the system which it is intended to supply. The transformer primary windings are permanently connected to the generator output terminals, without any switchgear or protective equipment between them and for practical purposes the transformer is regarded as part of the generator. The transformer windings may be delta/star or star/interstar and the neutrals of the generator and of the secondary windings are earthed in the usual way.

When stepping down to feed a low voltage system a delta/star configuration will usually be used. An interstar secondary winding has a high impedance to zero sequence currents and is not therefore suitable to feed a low voltage distribution system. If stepping up to feed a high voltage system a delta/star or star/interstar configuration may be used.

The high zero sequence impedance of the interstar secondary winding is then an advantage in that it limits earth fault currents. If the standby supply is being added to an existing installation, the effect of the high zero sequence impedance of the interstar secondary winding, on the operation of any earth leakage protection relays in the high voltage distribution system, should be considered.

When a generator transformer is fitted, the voltage regulator should, instead of setting the generator voltage, be arranged to set the transformer secondary voltage. This is best achieved by sensing the generator voltage and incorporating a volt drop compensation circuit within the voltage regulator. The compensation circuit includes an analogue of the transformer impedance and is supplied with an analogue of the load current. It is therefore able to calculate the voltage drop across the transformer and modify the generator voltage to maintain the correct voltage at the transformer secondary terminals. The alternative arrangement of sensing the transformer secondary voltage is not favored for two reasons. Firstly, a potential transformer may be necessary and secondly, the voltage signal will probably be carried over a long length of cable between the transformer and the voltage regulator. Such a cable is vulnerable to damage, and if the voltage signal is lost the generator has to be shut down.

Standard Reference Conditions

The standard conditions for ac generators are specified in ISO 8528 and BS EN 60034-1 as:

Altitude above sea level	1000 m
Cooling air temperature	40°C
Coolant temperature at inlet	25°C

Voltage Regulators

The voltage regulator maintains the output voltage of the generator within the specified limits. It has developed from a variety of ingenious electro-mechanical devices to the versatile electronic devices of today. The accurate and skilled assembly that the electro-mechanical regulators demanded is not compatible with modern manufacturing techniques, and friction and the inertia of moving parts set a limit to the accuracy and speed of response.

All regulators are now electronic, the final control being by transistors instead of variable resistors, which has led to improved accuracy and speed of response. Typical performance figures appear in ISO 8528 and are included in the section titled "Starting Mechanisms."

It is important to note that the voltage regulator on its own does not determine the recovery time of the generator output voltage, there are other and longer time constants in the system. The regulator can respond very quickly to a voltage dip by increasing the voltage across the exciter field circuit, but the exciter field circuit is inductive and introduces a time constant. The exciter in turn increases the voltage across the main field which is highly inductive and introduces an additional and longer time constant. The overall response time is improved by field forcing, or applying say twice the normal voltage to the exciter field thereby increasing the rate of rise of current in the main field.

For voltage rises the regulator can do no more than reduce the exciter field voltage to zero, leaving the main field current to circulate in the highly inductive loop formed by the field and the associated rectifier. The current will decay at a rate determined by the time constant which will be fairly long, but flux reduction is encouraged by the demagnetizing component of the load current (i.e., lagging vars).

Figure 1.6 shows a typical response curve for a voltage regulator.

There are four voltage regulating performance classes for diesel engines defined in ISO 8528 and the data relating to them appear in Table 1.1. These performance classes may be regarded as being equally applicable to gas turbine driven generating sets.

Speed Governors

The speed governor maintains the speed of the engine and generator within the specified limits. It has developed from the simple Watt governor to the sophisticated electronic governors available today.

The simple Watt governor has no amplifying mechanism and depends on the centrifugal force acting on two or more rotating weights, the force

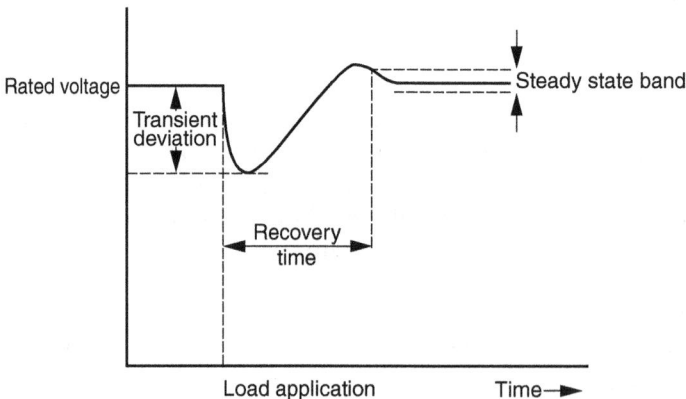

Figure 1.6 Typical voltage regulator characteristic (with zero droop).

TABLE 1.1   **Performance Classes for Diesel Engines**

Class	G1	G2	G3	G4
Frequency droop	8%	5%	3%	AMC
Frequency steady state band	2.5%	1.5%	0.5%	AMC
Transient frequency drop on application of maximum permitted step load	15% + droop	10% + droop	7% + droop	AMC
Transient frequency rise from initial frequency on loss of rated load	18%	12%	10%	AMC
Frequency recovery time	10 s	5 s	3 s	AMC
Voltage steady state band	±5%	±2.5%	±1%	AMC
Transient voltage drop on application of maximum permitted step load	25%	20%	15%	AMC
Transient voltage rise on loss of rated load	35%	25%	20%	AMC
Voltage recovery time	10 s	6 s	4 s	AMC

AMC = By agreement between manufacturer and customer.
Class G3 would be applicable to most standby power applications.

necessary to move the fuel rack depends on a speed error and there-fore introduces a load droop, and the response is slow. It is unlikely to be encountered on any modern engine or turbine.

The simple governor was improved by the addition of a hydraulic amplifier and a closed loop control system. The fuel rack is moved by the hydraulic system and the droop can be eliminated, the amplifier improves accuracy and speed of response, and load sensing can be introduced by the addition of a load measuring device to further increase the speed of response. With hydraulic governors load sharing between multiple sets is usually achieved by introducing a load-dependent droop.

Hydraulic governors have given excellent service for many years and many remain in use, but modern machines now use electronic governors. The electronic governor consists of three separate parts, the speed sensor, the electronic control system which may be mounted in the control cubicle or on the engine, and the fuel rack actuator which may be a sole-noid or a torque motor. The speed sensor may look at magnetic markers on the flywheel rim, at the starter ring, or at the frequency of the alter-nating generator output voltage. Looking at the generator output has the disadvantage that maloperation may be caused by a severely dis-torted wave form.

It is important that the electronic control system includes a built-in safety feature that ensures the set will be shut down if the speed sens-ing signal, for whatever reason, is lost.

Electronic governors make it easy to change the control parameters within the feedback loop and to send or receive electrical analogue signals. Functions such as load sharing, load management, and automatic synchronizing are easier than with hydraulic governors. Load sharing between paralleled sets is achieved by interconnecting a load share signal to each governor, if one machine is taking too great a share of load the sharing signal will react to reduce its fuel supply.

Early electronic governors used analogue systems which compared the measured quantity with the set value and made any necessary corrections to the fuel rack position. Modern governors use digital systems which digitize the measured and set values and feed the results into a processor which determines any corrections to be made to the fuel rack position. Digital systems are more versatile than analogue, changes to the operating parameters or sequence can often be achieved by a software adjustment instead of having to undertake a hardware modification. For the simplest applications digital systems have no particular advantages, but they may well become the normal manufacturing standard. Figure 1.7 is a block diagram showing the basic operating components of an electronic governor system.

There are four speed governor performance classes for diesel engines defined in ISO 8528 and the data relating to them appear in Table 1.1. It is important to note that the governor on its own does not determine the performance parameters, which are affected by the complete engine/generator system. The system includes the governor, voltage regulator, generator, flywheel, and the load and is affected by the surplus engine torque available for governing purposes.

Typical performance for gas turbines provides an accuracy on steady load within 0.5 percent and a transient deviation within ± 1 percent for

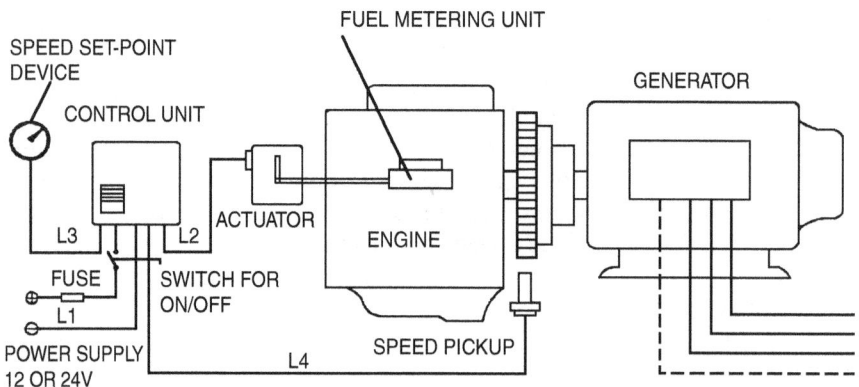

Figure 1.7 Block diagram of a typical electronic governor system. (Acknowledgements to Heinzmann United Kingdom Ltd.)

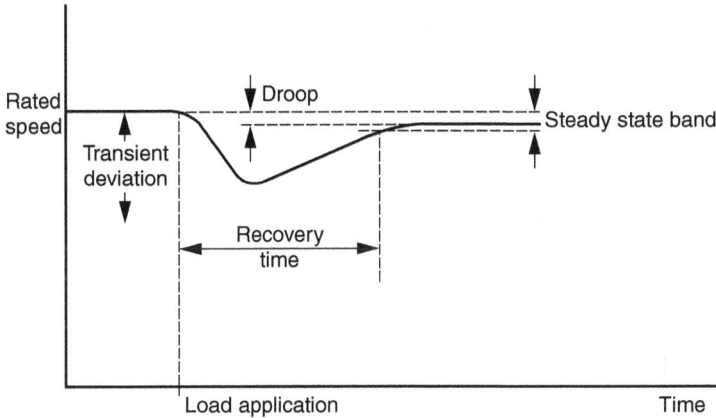

Figure 1.8   Typical speed governor characteristic (with droop).

2 seconds on application or removal of 25 percent rated load. Figure 1.8 shows a typical response curve for a speed governor.

Load-sensing speed governors increase the speed of response and reduce the speed dip. The electrical load is monitored and when a step increase is sensed a signal is passed to the governor control system which starts to move the fuel rack early, probably before the drop in speed is sensed.

Voltage and Frequency Performance Classes for Diesel Engine–Driven Sets

Specification ISO 8528 includes four performance classes for diesel engine–driven generating sets, the details of which appear in Table 1.1. Other performance classes are available, such as ISO 3046 (BS 5514) for speed governing and BS EN 60034 for voltage control but are not reproduced here.

Modes of Control

There are three modes of control which may be used in connection with speed governors and voltage regulators. The error is the basic feedback signal and is the difference between the measured value and the desired value.

■ Proportional control is the basic mode. Within a proportional band feedback is applied which is proportional to, and tends to reduce, the error. This form of control can achieve a stable state within the proportional band but there will be a deviation from the desired value, known as the offset.

- Integral control modifies the proportional feedback signal by adding to it a signal which has a rate of change proportional to the error, and attempts to nullify the offset. Adding integral to proportional control increases the accuracy at the expense of increasing the overall response time.

- Derivative control modifies the feedback signal by adding to it a signal which is proportional to the rate of change of the error. Adding derivative control to proportional plus integral control decreases the overall response time.

The initial letters P, I, and D are often used when referring to these modes of control. The proportional mode (P) will result in a speed or voltage droop. If isochronous running or constant voltage is required, integral control (I) is added to form a PI mode of control. Derivative control (D) may be added to the PI mode to form a PID mode of control which will result in an improved transient performance.

Starting Mechanisms

Deriving the Starting Power

Engine or turbine acceleration for starting will be by one or more battery driven electric motors or, for larger diesel engines only, by compressed air.

For diesel engines an electric starting system will include one or more preengaged electric starter motors engaging with a starter ring on the flywheel rim, a battery, and a battery charger. After a few revolutions the engine reaches its firing speed and develops sufficient power to accelerate itself to rated speed. For gas turbines a pony motor is mounted on the gearbox. It may take up to a minute for the rotor to reach its self-sustaining speed, and for the turbine to accelerate itself to rated speed and to be ready to accept load.

Compressed air starting is limited to diesel engines of ratings of the order of 1 MW or above; smaller engines do not have space in the cylinder heads for the air injection valves. Selected cylinders will be fitted with air injection valves and, during starting, a rotating air distributor valve feeds compressed air to appropriate cylinders in turn. Within the engine room there will be a compressed air cylinder with drain valve and pressure gauge, an electrically driven and automatically controlled air compressor to maintain the pressure within the vessel, and for black starts a diesel driven compressor. In the United Kingdom the compressed air installation will be subject to the Pressure Systems and Transportable Gas Containers Regulations 1989 as amended by S.I. 1991/2749. The regulations set standards for the installation and require the compressed air cylinder to be inspected at regular intervals by a competent person.

Failure to Start

If after cranking for the normal period of time the engine has failed to start, manufacturers adopt one of two philosophies. The cranking time may be extended to two or three times the normal period, or two or three successive starts may be attempted, with a short rest period between. The total period of time is limited by the battery capacity but another limitation is set by the accumulation of unburnt fuel in the cylinders, which is highly undesirable. The manufacturer's advice should be sought before any attempt is made to extend the cranking period.

Batteries and Battery Chargers

The starting battery will usually be a lead acid type and although the recombination type of lead acid battery is available for this application the high temperatures commonly encountered in engine rooms can lead to a short life and early battery failure. The flooded type of lead acid cell which can be topped up at regular intervals will survive the high temperatures and have a much longer life (provided it is topped up!).

The battery charger must be supplied from the distribution system which in turn is supplied either from the normal supply or from the standby supply, and should provide current-limited constant voltage charging. The old-fashioned trickle charging should not be used, it tends to cause unnecessary gassing and loss of electrolyte. The engine control system should provide visual indication and an alarm on charger failure. The starter battery and its charger are vital engine components and their neglect can lead to the failure of an engine to start.

Fuel Systems

Daily Service Tanks

Within the engine room there should be a daily service tank holding sufficient fuel for say 8 hours running; for a modern diesel engine 1 liter of fuel will provide approximately 3.5 kWh of electrical energy, for a gas turbine the fuel requirement will be greater, and 1 liter of fuel will provide approximately 2.2 kWh of electrical energy. The underbase of a diesel generating set is often used as a daily service tank; the arrangement requires a lift pump to provide a supply of fuel to the injection pump or pumps. The alternative is to mount a service tank at a level above that of the engine and to rely on gravity for the supply to the injection pump. For service tanks up to about 1000 liters the tank can be mounted on a stand fixed to the engine underbase, but larger tanks will for practical reasons be mounted on a floor-mounted stand. Fuel tanks and pipework should not be of galvanized iron as zinc and fuel oil are incompatible.

If extended running is required the daily service tank is normally topped up automatically, by gravity or by an electric pump, from a remote bulk fuel tank. Where there is no bulk fuel supply topping up of small or moderate-sized sets can be achieved manually using a semirotary hand pump attached to the engine base frame and provided with a flexible hose for extracting fuel from manually handled barrels at floor level. As an alternative a portable electric pump may be used but this would not be available for "black start" conditions. For larger sets arrangements can be made for fuel to be delivered by a small tanker vehicle; where this is intended a vehicle earthing facility should be provided to prevent any build up of a static electrical charge.

The daily service tank should be fitted with pipe connections for the fuel supply to the engine (the outlet pipe), the injector spill and excess fuel return from the engine, the fuel supply from the bulk tank, and a drain valve which may be plugged. There should be an easily visible contents gauge and, depending on the arrangement of the fuel system, a filler cap, an overflow connection, and a high-level alarm switch arranged to alert personnel and to shut down any transfer pumps. Within the tank the outlet pipe should finish above the base of the tank to prevent water and sediment from entering the fuel line.

In order to allow fuel to be extracted and supplied to the engine the daily service tank must be vented. If the supply to the daily service tank is automatically controlled as described in Figs. 1.9 or 1.10, the daily service tank is normally vented to the bulk tank which in turn is vented to atmosphere; there must be no shut-off valves in vent pipes. If the daily service tank is topped up manually, the tank must be provided with its own filler cap and vent.

Before the fuel enters the fuel pump it should pass through a fine filter which is invariably fitted to the engine. If long periods of running are envisaged, a dual filter arrangement can be provided; this enables a change of filter element to be undertaken while the engine is running.

Bulk Fuel Tanks

A detailed description of the bulk fuel tank installation is beyond the scope of this book but the following comments may be helpful.

If the bulk tank is mounted below the daily service tank, Fig. 1.9 indicates the basic interconnecting requirements. An electrically driven

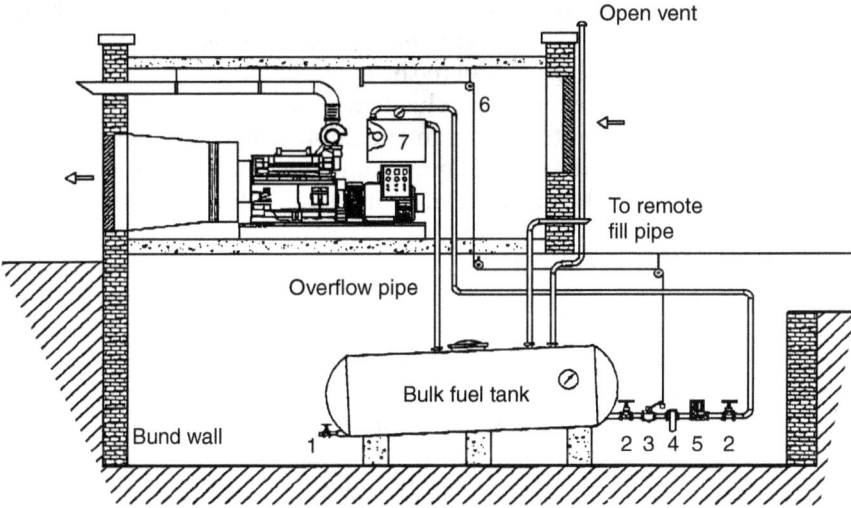

Open vent

6

7

To remote
fill pipe

Overflow pipe

Bulk fuel tank

Bund wall

1

2 3 4 5 2

Key:
1. Sludge cock
2. Isolating valve
3. Weight operated fire valve
4. Strainer
5. Transfer pump
6. Fusible link
7. Daily service tank fitted with
   high level switch and alarm

Figure 1.9 Fuel system, pumped feed from a bulk fuel tank. (Acknowledgements to the Association of Manufacturers of Power Generating Systems.)

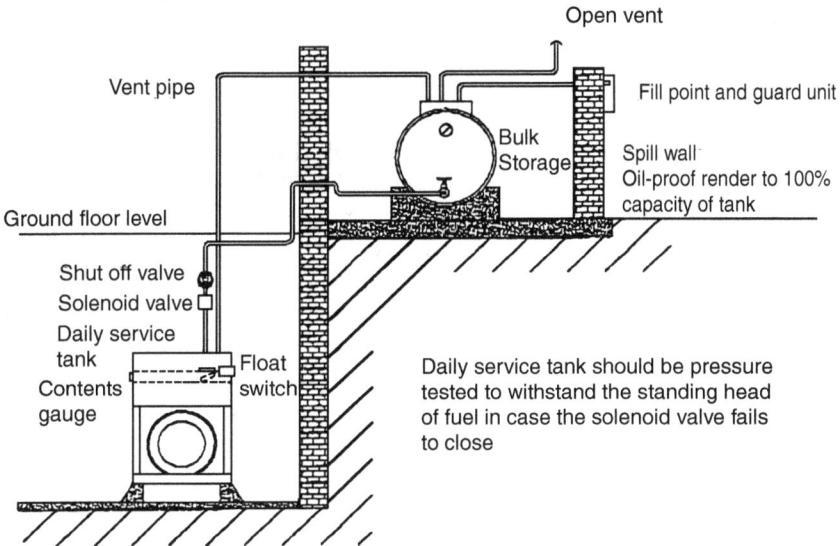

Open vent

Vent pipe

Fill point and guard unit

Bulk
Storage

Spill wall
Oil-proof render to 100%
capacity of tank

Ground floor level

Shut off valve
Solenoid valve
Daily service
tank
Contents
gauge

Float
switch

Daily service tank should be pressure
tested to withstand the standing head
of fuel in case the solenoid valve fails
to close

Figure 1.10 Fuel system, gravity feed from a bulk fuel tank. (Acknowledgements to the Association of Manufacturers of Power Generating Systems.)

pump is installed between the daily service tank and the bulk tank, the pump being controlled from start/stop level controls in the daily service tank. To avoid spillage of fuel in the event of failure of the level controls, an overflow pipe returns surplus fuel back to the bulk tank. There should be a manual shutoff valve in the supply pipe to allow for occasional maintenance of the electric pump and any other items installed in the supply pipeline. The daily service tank should have a filler cap and vent, the filler cap is available for emergency use if the pump fails, or if the bulk tank is empty or not available for use. It is most important that the overflow pipe be large enough to allow the free flow of the maximum quantity of fuel that can be delivered by the pump. This should be demonstrated during the commissioning procedure.

If the bulk tank is mounted above the daily service tank, Fig. 1.10 indicates the basic interconnecting requirements. The fuel level within the service tank is maintained by a solenoid valve in the supply pipe run and open/close level controls within the service tank. The service tank and the associated pipework should be pressure tested for the maximum static pressure that can arise if the level controls or the solenoid valve fail. A vent pipe from the daily service tank is taken back to the bulk tank, or is vented to atmosphere above the level of the bulk tank; it follows that there cannot be a spillage of fuel from the service tank unless there is a failure of the pipework or of the tank. There should be a manual shutoff valve in the supply pipe to allow for occasional maintenance of the daily service tank and any items installed in the supply pipeline.

Although reliance on gravity may appear to simplify the installation, this is not the preferred arrangement. The failure of the solenoid valve results in the service tank being subject to the full static pressure from the bulk tank. Even a slow leakage through the valve can, between test runs, result in the service tank being full and the vent pipe being full of fuel to the level of the bulk tank. For these reasons a filler point cannot be allowed on the service tank, there is no way of ensuring that it is properly replaced. It follows that if there is only one bulk tank and it is empty or out of use, manually filling the service tank presents a problem. Temporary arrangements will normally have to be made within the engine room, alternatively permanent pipework could link the service tank to a remote fill point at the same level as the bulk tank. Remote filling requires a communication link between the fill point and the engine room.

Bulk storage tanks are usually cylindrical, mounted horizontally with a slight incline, fuel being taken off from the higher end while a water drain point or sludge cock at the lower end allows water and sludge to be drawn off. There should be local or remote indication of the contents, a filling point and an access manhole, raised above the tank to allow entry (strictly controlled) for occasional cleaning or inspection. At the fill point there should be an audible warning which sounds when the bulk tank is full.

When fuel is delivered to the bulk tank, care has to be taken to avoid introducing foreign material into the tank and there should be arrangements for earthing the delivery vehicle to avoid a discharge of static electricity during delivery. A coarse filter should be installed in the pipeline to the service tank, before the electric pump or any other item in the pipeline.

Fire Precautions

A risk assessment should be undertaken which considers the results of an engine room fire. Local regulations regarding the storage of fuel vary and at an early planning stage the installation should be discussed with the local planning and fire authorities so that the installation can be planned for compliance. In the United Kingdom the planning authority is the first point of contact, they will effect an introduction to the fire authority. The fire authorities are mostly concerned with escape routes from the building, but the occupier should consider wider implications such as the potential loss of the building and its contents, the loss of business records, and the loss of profit due to the inability to operate after a fire.

Where the daily service tank is supplied by gravity from a bulk tank there is usually a requirement to install a fire valve, closed on receipt of an alarm indicating a fire within the engine room, between the bulk fuel tank and the engine room. This prevents fuel entering the area and feeding the fire.

If the generating set is installed within an occupied building there may be a requirement to drain the daily service tank within say 1 min of receipt of an alarm indicating a fire within the engine room. This requires a jettison pipe connection to the daily service tank and a jettison valve opened on receipt of a fire alarm. If the bulk tank is below the service tank fuel can be returned to it by gravity, but if not, a jettison tank (normally empty) outside of the engine room and below the level of the daily service tank will be required. The jettison tank will require an air vent which may be piped back to the service tank or may be terminated in a safe area. There must be clear indication, in an appropriate part of the building, of any liquid within the jettison tank and procedures must ensure that it is emptied without delay. A method of emptying the jettison tank should be available, if no drain cock is available this could be by using a portable electric or semi-rotary hand pump.

Fire valves are usually of the free-fall gravity-operated type, electrical power for operation may not be available when required. Within the engine room there should be a reliable fire detection system not subject to spurious operation by the oil or exhaust fumes that are always present, or by the high ambient temperatures that are sometimes experienced. Suitable fire detection systems are available, one that has been in use for many years uses fusible metal links mounted above the engine and flexible cables running around pulleys and extending to a fire alarm switch, and to any local free-fall fire valves. Whatever fire alarm system is installed in the engine room, it should be interconnected with any main building fire alarm system.

In the United Kingdom there is a growing requirement for service tanks to be double skinned or bunded, and for fuel pipework to be double skinned or clad in such a manner that any leakage will fall under gravity to a bunded area. There must be indication that a double-skinned tank has failed, and indication of any liquid in a bunded area. This implies that there must be a responsible person who will receive the information and has the authority to take appropriate action.

Provided that the engine room is kept clean and is not used as a storage area for miscellaneous combustible material, the usual cause of fire within a diesel engine room is the fracture of a high pressure (injector) pipe suffering from vibration and fatigue. The fracture results in fuel being finely sprayed in an indeterminate direction. Diesel fuel is not easily ignited but if the spray is directed towards a hot turbocharger or exhaust manifold fire is likely to result. This is a hazard more likely to be experienced with continuously running engines, standby power generating sets do not usually build up sufficient running hours to suffer from fatigue failures.

Waxing of Fuel at Low Temperatures

At low temperatures the wax content of diesel fuel separates out as fine floating particles and in order to avoid problems with road vehicles the oil companies distribute different grades of fuel for use during summer and winter months. For standby installations it is not possible to predict when the fuel will be used or to purchase winter grade fuel in summer months. It therefore has to be assumed that, in the United Kingdom, the fuel stored is summer grade and liable to waxing during winter. The waxing temperatures are 0°C for summer grade and −9°C for winter grade. The waxing does not prevent the flow of fuel through pipework but causes filters and injectors to be clogged.

If the air temperature within the engine room is maintained at a few degrees above 0°C, the fuel in the daily service tank and in the engine fuel pipework will not be subject to waxing. During running, cold fuel supplied to the service tank will mix with the warm fuel already there and any wax particles will be dissipated.

If the temperature within the engine room is not maintained above 0°C trace heating of the engine pipework and filters should be considered, or discuss with the fuel supplier the possibility of using an antiwaxing additive. There are practical difficulties in ensuring that additives are dispersed throughout the store of fuel.

If the bulk tank is installed outside of the building, the filter at the outlet should be a coarse filter that will not clog under cold conditions, otherwise some trace heating of the filter pipework run will be required.

Long-Term Storage of Diesel Fuel

If a strategic quantity of fuel is stored on site for emergency purposes, there are two hazards which may be encountered, microbiological growth and chemical degradation. Water will inevitably collect at the base of the tank due to internal condensation and microbiological growth occurs at the oil/water interface. The water should be drained off as necessary and any significant biological growth will be observed as sludge. If sludge is observed specialist attention may be required. It is sometimes recommended that 25 percent of the contents of a tank should be used and replaced with new fuel each year.

Chemical degradation occurs independently of microbiological growth and results in a darkening of the color of the fuel.

The problems are best avoided by good housekeeping involving regular draining of water from the tank, regular testing of the contents by an experienced specialist organization, and regular usage so that new oil can be added. These remarks are mainly directed at bulk tanks, but water (and any sludge) should be regularly drained from service tanks.

Engine Cooling Systems

Water-Cooled Diesel Engines

Up to 40 percent of the energy of the fuel used by a modern diesel engine is transferred to the crankshaft, the remaining energy is dispersed from the exhaust gases, the radiator, and the engine casing. Manufacturers' data for these losses vary widely but guideline figures may be considered as:

Useful energy transferred to generator	38.5 percent
Energy used by engine driven radiator fan	1.5 percent
Heat lost in exhaust gases	30.0 percent
Heat lost from radiator	25.0 percent
Heat lost from engine casing	5.0 percent

If the radiator fan is not engine driven, an equivalent amount of additional power will be available from the generator output where it will be required to supply an electrically driven radiator fan.

For sets rated up to about 800 kW the radiator and its associated cooling fan are usually mounted, on the base frame, at the nondriving end of the engine, the fan being belt driven from the crankshaft. Within the engine room the set may then be positioned perpendicular to a perimeter wall which is provided with an opening matched in size to the radiator. This arrangement is convenient and ensures that the cooling air passes over the engine in an efficient manner. To the purist it has the slight disadvantage that the cooling air is extracted from the engine room and will be above the outside ambient temperature and hence the radiator is not used in the most efficient manner.

For larger sets the radiator is usually mounted outdoors in a horizontal or vertical plane and uses one or more thermostatically controlled, electrically driven, forced-draft fans. Horizontally mounted radiators have the advantage of directing the warmed air upward where it is less likely to cause a nuisance to personnel.

Whether the radiator is mounted in the engine room wall or outside of the building the noise produced by the fan or fans will require consideration and acoustic attenuators may be required. Wherever hot water pipework is within reach of personnel, inside or outside of the engine room, it is a source of danger and must be thermally insulated to reduce the surface temperature to a safe value.

Lubricating-oil cooling is variously achieved: an oil/water or an oil/air heat exchanger may be used or a section of the local radiator may be dedicated to oil cooling.

Direct Air–Cooled Diesel Engines

Up to ratings of the order of 100 kW direct air–cooled diesel engines are available. For this arrangement the cylinders are provided with a cladding having air inlet and outlet flanges. A fan forces air into the spaces between the cylinders and the cladding where it absorbs the engine's rejected heat and is forced to the outlet opening. The route of the air flow is necessarily contorted which leads to a large pressure drop across the air flow passages and the air outlet flange will usually be much larger than the inlet flange.

Within the engine room the set should be positioned so that the exiting cooling air may be ducted directly to the exterior of the building by a straight length of ducting of adequate cross section, say 125 percent of the outlet flange area. The engine will normally be provided with an engine-driven fan with some surplus power, but if any significant inlet or outlet ducting, louvers or acoustic attenuators are to be installed, either a more powerful fan or an additional fan will be required.

Direct air–cooled diesel engines are simpler than radiator-cooled engines: there are no corrosion or freezing problems and no leaking from failing pipe work or from burst hoses.

In the preceding text some guideline figures are given for water-cooled engine losses and efficiency. If, for simplicity, it is assumed that the efficiency, the casing losses, and the exhaust losses of direct air cooled engines are similar, the allocation of losses for direct air cooled sets may be restated as:

Useful energy transferred to generator	38 percent
Energy used by engine driven cooling fan	2 percent
Heat lost in exhaust gases	30 percent
Heat lost to cooling air	25 percent
Heat lost from engine casing	5 percent

Gas Turbine Cooling

A gas turbine will have smaller losses from the turbine casing and there is no radiator, most of the waste heat is in the exhaust gases. Because of the high rotor speed the casing of a turbine is much smaller than that of the equivalent engine and it is more easily thermally insulated; this leads to the smaller casing losses. Heat is removed from the rotor bearings by lubricating oil flow and the lubricating oil system will include an oil cooler, probably of the oil/air type discharging air to outside of the building.

Engine Room Ventilation

The temperature within the engine room must be maintained at a reasonable level, and the heat losses from the engine and generator have to be removed by a flow of ventilation air. The heat losses will be mainly from the engine, the losses from the generator will be only a few percent (typically 5 percent) of its rating; the heat losses from a turbine and its gearbox will be much less than from a diesel engine of comparable rating.

The quantity of air entering the room will always exceed the quantity leaving; combustion air is drawn into the engine and is dispersed in the exhaust gases. The combustion air requirement is small and for diesel engines may be assumed to be 0.17 m³/s for a 100 kW set and pro rata for other ratings. A gas turbine requires more, typically twice as much as a diesel engine (3.4 m³/s for a 1 MW set).

Water-Cooled Diesel Engines

It is necessary to estimate the heat losses into the engine room; these will originate from the engine casing, the generator, and the exhaust system. The generator efficiency will be of the order of 95 percent and the heat balance figures from the previous section may be rearranged as:

Useful energy (kWe) appearing at the generator terminals	36.5 percent
Engine-driven radiator fan	1.5 percent
Heat lost in exhaust gases	30 percent
Heat lost from radiator	25 percent
Heat lost from engine casing	5 percent
Generator losses	2 percent

Estimated values for the engine casing and generator losses may be derived from the above heat balance. The exhaust system losses will vary widely depending on the length of internal pipework, whether there is an internal silencer, and the amount of thermal insulation. In the following example a loss of 6 percent of the generating set kWe is assumed from the exhaust system.

If the generating set kWe power rating is indicated by P the losses may be listed:

- Losses from the engine casing $(0.05 \div 0.365)$ $0.137 \times P$ kW
- Losses from the generator (assumed to be 5%) $0.05 \times P$ kW
- Losses from the exhaust system (assumed to be 6%) $0.06 \times P$ kW
- Total losses to room $0.247 \times P$ kW

The standard reference conditions specified in ISO 8528 allow a maximum air temperature of 25°C and knowing the losses into the engine room, and the permissible rise in temperature of the ventilating air, it is possible to calculate the air flow required from the formulae:

$$\text{Air mass flow (kg/s)} = \frac{\text{Heat losses into engine room (kW)}}{\text{Temperature rise (°K)} \times \text{Specific heat capacity of air (kJ/kg °K)}} \quad (1.2)$$

$$\text{Air volume flow (m}^3\text{/s)} = \frac{\text{Air mass flow (kg/s)}}{\text{Density of air (kg/m}^3\text{)}} \quad (1.3)$$

As an example the probable ventilation requirements for a 100 kW generating set may be estimated in the following fashion. (For the final planning or design stage it is necessary to use the heat loss figures from the manufacturers or installers.)

Assume that the entering air is at 15°C and a 10°K rise is permissible, the specific heat capacity of air may be taken as 1.005 kJ/kg°K, and the density as 1.22 kg/m³.

From Eq. 1.2

$$\text{Air mass flow} = \frac{100 \text{ kW} \times 0.247}{10°\text{K} \times 1.005 \text{ kJ/kg}°\text{K}}$$

$$= 2.46 \text{ kg/second}$$

From Eq. 1.3

$$\text{Air volume flow} = 2.01 \text{ m}^3/\text{second}$$

Where the radiator is mounted within the engine room and takes its cooling air from within the room, it will invariably be the case that the radiator air flow exceeds the room ventilation requirements. The standard radiator fan fitted to an engine will have some surplus power and will probably overcome a total external pressure drop of 250 Pa (2.5 mbar). However, if room inlet or outlet louvers or acoustic attenuators are to be fitted, the total external pressure drop should be ascertained and the set manufacturer consulted. If necessary a supply fan should be installed. Figure 1.11 indicates a typical arrangement for a local radiator cooled diesel engine driven generating set.

Where the radiator is mounted away from the engine, engine room ventilation will be essential and Fig. 1.12 indicates a typical arrangement for a remote radiator-cooled diesel engine–driven generating set.

Direct Air–Cooled Diesel Engines

Air is extracted from the engine room to cool the engine but this will not be sufficient to remove the heat gained from the engine cladding. Engine room ventilation will be required as discussed in the preceding paragraphs; Figure 1.13 indicates a typical arrangement for a direct air–cooled diesel engine–driven generating set.

Gas Turbine Ventilation

The ventilation requirements for a gas turbine are calculated in a manner similar to that discussed in the preceding paragraphs. The heat loss data will be required from the manufacturer. Turbines are sensitive to any restrictions of flow of the air inlet or of the exhaust outlet

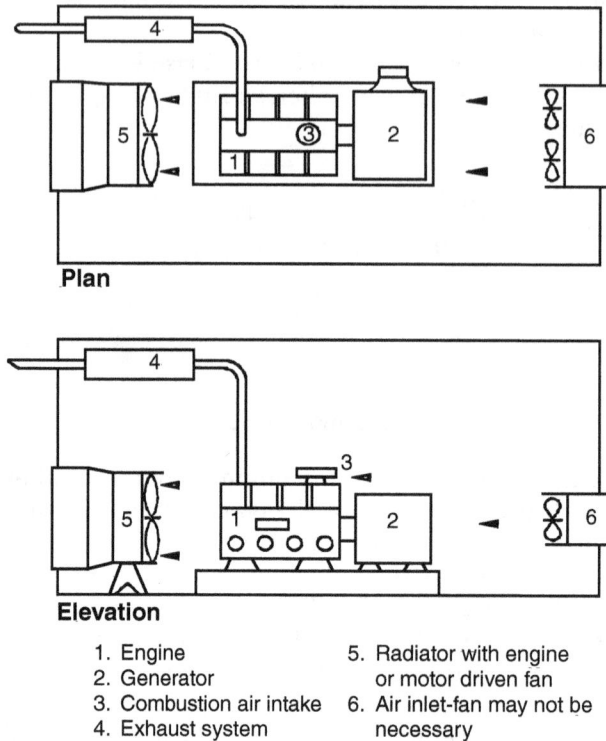

Plan

Elevation

1. Engine
2. Generator
3. Combustion air intake
4. Exhaust system

5. Radiator with engine
   or motor driven fan
6. Air inlet-fan may not be
   necessary

Figure 1.11   Typical arrangement for a local radiator-cooled diesel engine–driven generating set.

and the arrangements for turbine air inlet and exhaust are much larger than for a diesel engine, the layout will be constrained by the turbine design. Figure 1.14 indicates a typical arrangement for a gas turbine–driven generating set.

Direction of Airflow

The airflow within the engine room should preferably be arranged to be in the direction from the generator to the engine and radiator; this allows the generator to receive a supply of clean cool air. The generator produces only a small amount of heat and is cooled by a flow of cooling air which exits as clean warm air. The engine on the other hand produces a large amount of heat and various surfaces tend to produce oily vapor which would adversely affect the generator windings.

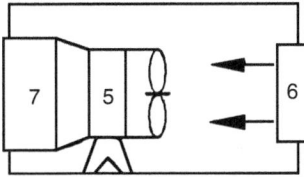

Elevation B-B

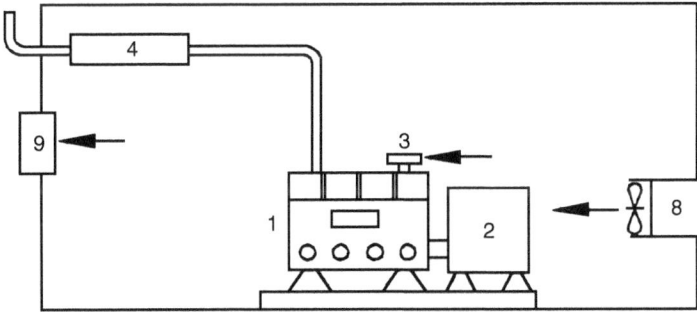

Elevation A-A

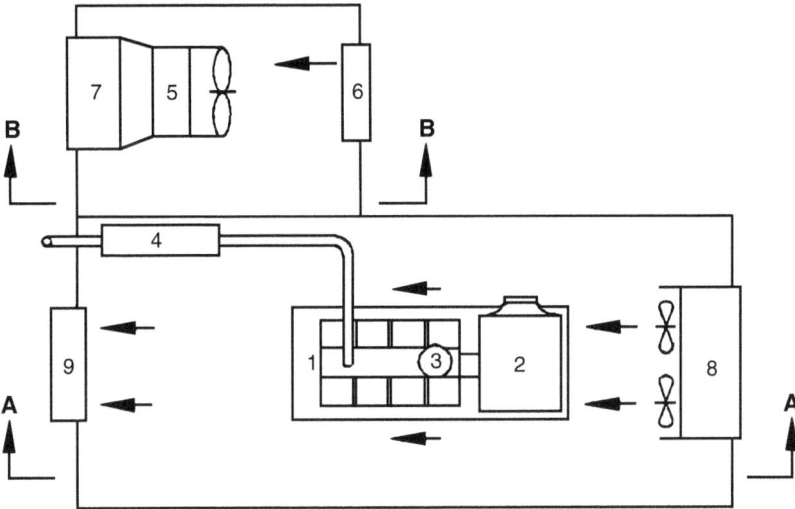

Plan

1. Engine
2. Generator
3. Combustion air intake
4. Exhaust system
5. Radiator with motor driven fan

6. Radiator air inlet
7. Radiator air outlet
8. Ventilation air inlet
9. Ventilation air outlet

Figure 1.12 Typical arrangement for a remote radiator-cooled diesel engine–driven generating set.

ELEVATION A-A

PLAN

1. Engine
2. Generator
3. Combustion air intake
4. Exhaust system
5. Engine cooling air inlet
6. Engine cooling air outlet
7. Flexible duct section
8. Cooling and ventilation air inlet
9. Ventilation air outlet

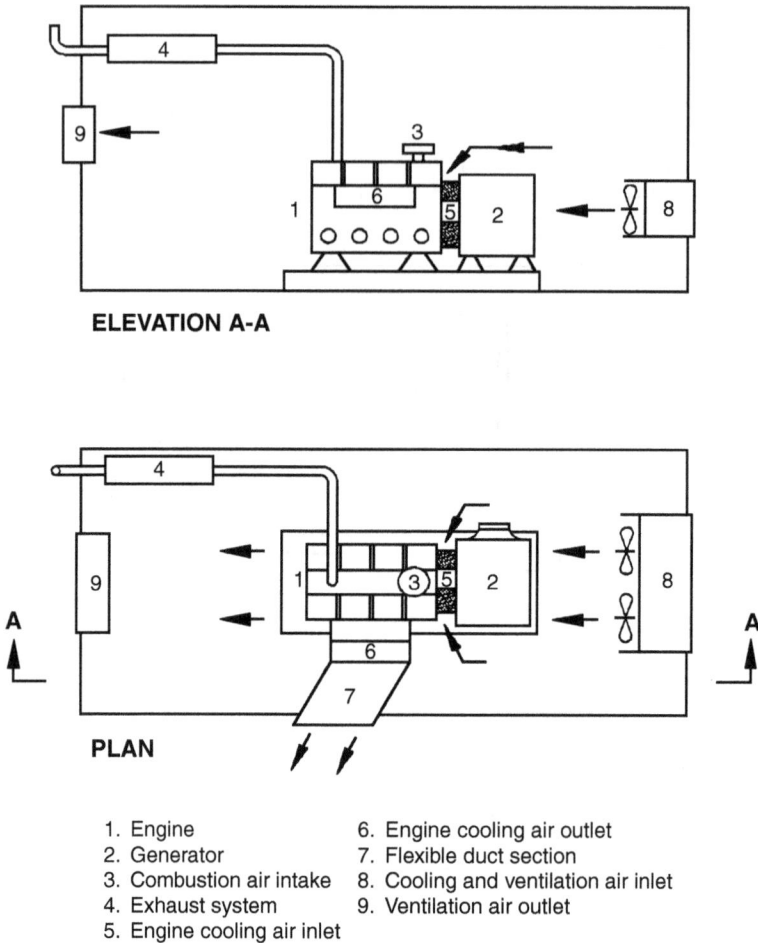

Figure 1.13 Typical arrangement for a direct air–cooled diesel engine–driven generating set.

Inlet and Outlet Louvers

A small amount of heating should be provided in the engine room to ensure that the temperature does not drop below say 5°C. Pivoted rain-proof louvers are therefore fitted to air inlet and outlet openings, arranged to be closed during idle periods and open while running, the louvers being opened as part of the starting sequence.

A common arrangement is for the louvers to be opened by spring or gravity and closed by electric motor, solenoid, or compressed air. This avoids the need for electrical power during the starting sequence when the normal supply has probably been lost. The arrangement can be reversed whereby the louvers are power opened and spring or gravity

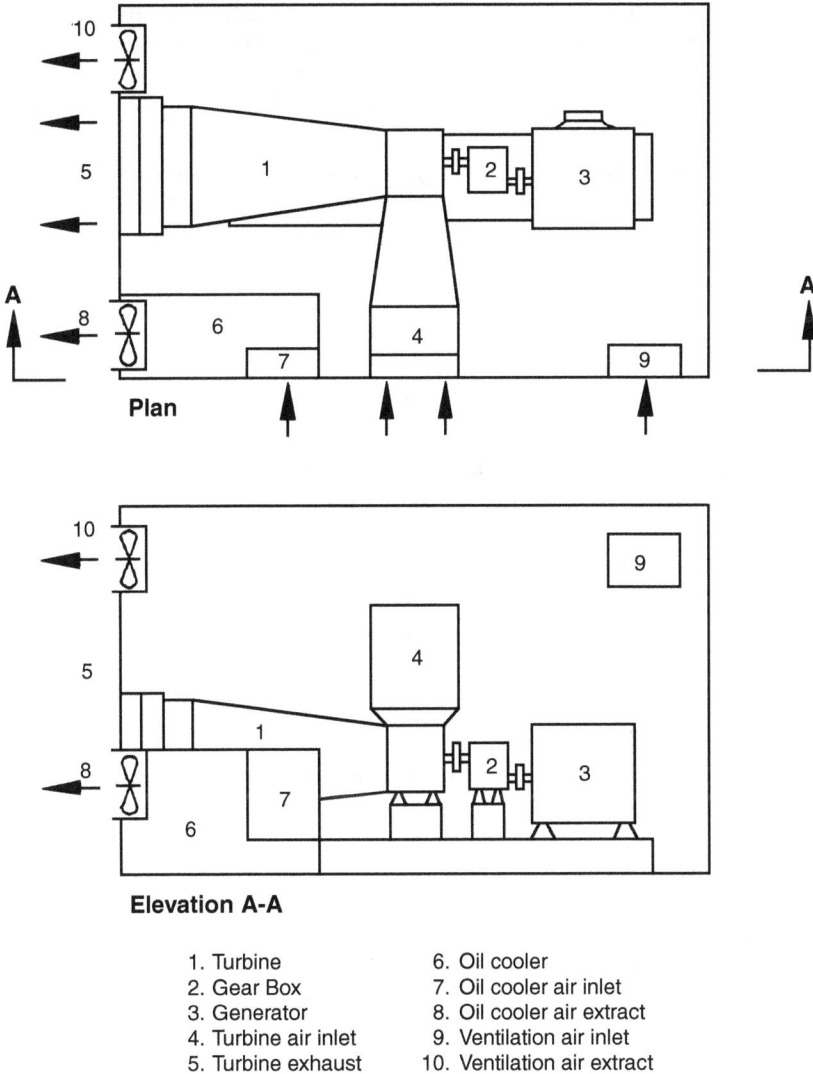

Figure 1.14 Typical arrangement for a gas turbine–driven generating set.

closed; in such cases the power opening will usually use the dc supply from the starting battery. If the engine starts with all louvers closed and the engine room is "sealed," the loss of combustion air will cause a reduction of pressure; to guard against this, electrical interlocks may be provided to ensure that the louvers are open before the starting sequence commences or a small area of inlet louver may be balanced, in other words, opened by a pressure difference.

The openings should be protected by a wire mesh to prevent the entry of birds or other wild life; if the mesh is on the outside of the louvers, birds are prevented from nesting and, if the openings are at street level, persons passing by are prevented from depositing litter within them.

The pressure drops across the inlet and outlet louvers must be taken into account when calculating the fan duties, if acoustic louvers are fitted the pressure drops will be higher and the effect on the fan duties will be greater.

Exhaust Systems

Gas Turbines

It has already been stated that the layout of gas turbines is constrained by the design of the air inlet and exhaust outlet arrangements. Gas turbines discharge their exhaust to atmosphere; there is no system of exhaust pipework and silencers as there is with diesel engines. The requirements of the Clean Air Act do not apply to gas turbines, which require special consideration by the local authority.

For these reasons the remainder of this section applies only to diesel engine exhaust systems.

Within the Diesel Engine Room

Pipe work will be fixed within the engine room to carry the exhaust gases from the engine outlet flange, through the wall or ceiling, to the outside of the building. The pipe work is fixed to the building structure, either from hangers fixed to the walls or ceiling or supported from the floor; to allow for the inevitable movement of the engine on its vibration dampers, a flexible section of pipe is provided between the engine outlet flange and the fixed pipe work. Where the pipe passes through walls or ceilings, precautions should be taken to avoid cracking of the structure due to heat and vibration. A sleeve or wall plates with clearance holes should be considered.

The exhaust gas temperature at the engine outlet will be of the order of 600°C and the exhaust pipe work will be subject to thermal expansion. Expansion is usually permitted by incorporating stainless steel bellows type couplings in pipe runs. Wherever the pipe is within reach of personnel, inside or outside of the engine room, it is a source of danger and must be thermally insulated to reduce the surface temperature to a safe value. To reduce the heat radiated within the engine room the entire internal exhaust system may be thermally lagged, this also reduces the noise transmitted to the room. Both these features make the working environment more tolerable.

Exhaust Silencers

The open outlet end of the exhaust system is a source of considerable noise, a prominent frequency being the (low-frequency) engine firing rate, and depending on the acoustic limitations of the site, one or two silencers will be included in the system, the primary silencer being mounted close to the engine while the secondary may be outside of the engine room.

If one silencer only is installed it will probably include reactive and absorptive sections, if two silencers are used, the primary silencer is likely to be reactive for attenuation of the lower frequencies, and the secondary absorptive for attenuation of the higher frequencies. Standard industrial silencers are available which provide an attenuation between 20 and 40 dB(A) at the exhaust outlet.

Discharge of Exhaust Gases

To comply with the Clean Air Act Memorandum (see the next subsection) the discharge of exhaust gases should be vertical, and to prevent the entry of rain it is usual to terminate the pipe obliquely and to provide a gravity-closed weather flap that is opened by the flow of the exhaust gases. If the end is open and is likely to allow excessive entry of rain, leading to internal corrosion, it would be prudent to provide one or more drain plugs at suitable points within the engine room. Test runs would boil off this water but if test runs are expected to be infrequent, provision should be made for drain plugs.

The open outlet of the exhaust system is a source of noise and noxious gases, its location has important environmental consequences and demands some attention at the planning stage. The following criteria should be taken into account:

- The minimum height of the outlet will be determined by The 1956 and 1993 Clean Air acts. The actual height of the outlet is subject to the approval of the local planning authority. The Clean Air acts are discussed in the next subsection

- The outlet should not be near any inlet grilles or opening windows that would allow the gases to enter any building.

- Exhaust backpressure reduces the engine output; the pipe run should be as short as practicable and the number of bends should be kept to the minimum. Bends should be of long radius type. If the length exceeds 10 m a larger bore may be necessary.

- The location of the outlet should take into account the effect of both noise and gases on the occupants of any nearby buildings, which may be offices, commercial or industrial premises, public buildings, or residential premises.

Generating sets are sometimes installed at ground level in light wells enclosed by buildings. In such circumstances there is no alternative to running the exhaust pipe vertically and terminating it, in accordance with the Clean Air Acts requirements, above the highest part of the building. This may result in a pipework run much in excess of the 10 m mentioned earlier, and in such cases it is essential that the engine manufacturer should be consulted. The engine performance will be affected and some adjustments may be required.

The 1956 and 1993 Clean Air Acts

The Clean Air acts apply to engine exhaust gases and the requirements are set out in the Third Edition of the 1956 Clean Air Act Memorandum published by Her Majesty's Stationery Office. The acts seek to ensure adequate dispersal of sulphur dioxide and other pollutants produced in normal combustion.

The acts do not apply to gas turbines which require special consideration by the local authority, nor do they apply to plants with a gross heat input of less than 150 kW. If the efficiency of the generating set is taken to be, say 35 percent, then they will not apply to installations of less than 52.5 kW rating. If it is not clear whether the acts apply or not, the gross heat input at maximum rated load, including any overload, should be obtained from the generating set manufacturer.

The Memorandum describes the method of calculating the minimum height of the exhaust gas discharge. There is a recommended minimum efflux velocity of 6 m/s but this is unlikely to be a problem for diesel engine exhausts; it is primarily applied to boiler flues and demands a vertical discharge. In determining the minimum height, the Memorandum takes into account three factors: the rate of emission of sulphur dioxide at the maximum power output including any overload capacity, the type of area in which the installation is situated, and the effect of adjacent buildings on flue gas dispersal.

The first two factors are used to determine, in the words of the Act, the uncorrected chimney height, and the third factor is used to modify this to arrive at the corrected chimney height.

The rate of emission of sulphur dioxide depends on the sulphur content of the fuel, and the maximum rate of fuel consumption. For BS 2869 Class A2 fuel (Class A1 fuel is no longer listed) the sulphur content should not exceed 0.2 percent by mass, and the rate of fuel consumption will depend on the rating and overall efficiency of the generating set and should be ascertained from the manufacturer. If it is proposed to use other fuels the sulphur content should be ascertained from the supplier. The atomic weights of sulphur and of oxygen are 32 and 16, respectively; it follows that for every kilogram of sulphur burnt, 2 kilograms of sulphur dioxide will be produced.

The types of area considered in the Memorandum are:

A. An undeveloped area where development is unlikely
B. A partially developed area with scattered houses
C. A built-up residential area
D. An urban area of mixed industrial and residential development
E. A large city or an urban area, of mixed heavy industrial and dense residential development

Having calculated the uncorrected chimney height, the effect of nearby buildings has to be considered, and the Memorandum describes the method to be used to establish the corrected chimney height. The correction procedure applies to all buildings within a distance of five times the uncorrected height, and to the building housing the generating set itself.

Clause 25 of the Memorandum introduces three additional requirements which are:

a. A chimney should terminate at least 3 m above the level of any adjacent area to which there is general access.
b. A chimney should never be less than the calculated uncorrected chimney height.
c. A chimney should never be less than the height of any part of an attached building within a distance of five times the uncorrected chimney height.

The heights determined by these procedures should be rounded to the nearest meter, and should be regarded as a guide rather than a mathematically precise statement. The conclusions may need to be modified in the light of particular local circumstances such as topographical features.

Example of Chimney Height Calculation

The manufacturer of a 180-kW generating set running on BS2869 Class A2 fuel advises that it uses 47 kg of fuel per hour when running at full load.

The maximum permissible sulphur content (0.2%) of 47 kg of fuel is 0.094 kg which when burnt will result in the emission of 0.188 kg/hour of sulphur dioxide.

Referring to Chart II of the Memorandum, the following uncorrected chimney heights are obtained:

For Type A district   1.9 m

For Type C district   2.6 m

For Type E district   3.3 m

The uncorrected height so obtained has to be corrected for the effect of nearby buildings. The method is clearly set out in the Memorandum with illustrations and examples. The heights determined by these procedures should be rounded to the nearest meter, and should be regarded as a guide rather than a mathematically precise statement. The conclusions may need to be modified in the light of particular local circumstances such as topographical features.

Pollutants Included in Exhaust Discharge

Exhaust gases discharged from the engine include, in addition to carbon dioxide (CO_2) and water vapor (H_2O), carbon monoxide (CO), various nitrous oxides (NO_X), aldehydes (HCO), unburnt hydrocarbon fuel (HC), and particulate matter. There will also be a small quantity of sulphur dioxide, depending on the sulphur content of the fuel. Carbon dioxide, water vapor, and any sulphur dioxide are the natural result of burning hydrocarbon fuel and cannot be reduced. Carbon monoxide and aldehydes are the result of incomplete combustion, nitrous oxides are formed by hot spots within the combustion spaces, unburnt fuel is fuel which has escaped the burning process, and particulate matter is finely divided matter, mainly carbon, resulting from incomplete combustion of the atomized fuel injected into the combustion spaces. All these pollutants are being reduced as the technology of injectors and combustion spaces advances.

These pollutants are now controlled by European legislation, on a gram per kWh basis, for road vehicle engines. As the efficiencies of engines are increased the pollutants are reduced but legislation may appear which relates to standby installations and installers should be aware of any new legislation. Engine and generating set makers will of course be aware of the latest requirements.

Catalytic Converters

Catalytic converters are installed in line with a run of diesel engine exhaust pipe in the same way as silencers; they include a ceramic honeycomb providing an enormous surface area for a catalytic metallic film over which the exhaust gases pass. They increase the backpressure of the exhaust system and therefore reduce, by a few percent, the efficiency of the engine; it follows that their use is not without an environmental "disbenefit." With the increasing attention being given to health and safety it may not be long before catalytic converters are required on some standby sets.

There are two types of catalytic converter: first, the two-way or oxidation converter, and second, the three-way or oxidation-reduction converter. The former uses a platinum and palladium catalyst which oxidizes carbon monoxide and hydrocarbons to form carbon dioxide and water, using surplus oxygen in the exhaust gases to do so. The latter uses a platinum and rhodium catalyst which, in addition to oxidizing carbon monoxide and hydrocarbons, reduces nitrous oxides to nitrogen and oxygen. It requires a stoichometric air/fuel ratio for proper operation and is not at present applicable to diesel engines used for standby generation purposes.

Catalytic converters do not operate until heated to their operating temperature, which is of the order of 200–300°C. Converters cannot change the particulate material in the exhaust gases but it has been claimed that by oxidizing the sticky hydrocarbons associated with it, the particulate material is changed to a dry powder which is much less damaging environmentally. This feature is useful as it offers the advantage that the particulate material adheres only temporarily to the catalytic surfaces.

Control Systems

The control system coordinates the operation of the various controls built into the generating set. Over many years the control system has developed from an assembly of hard-wired relays, indicator lamps, and meters to a small modular system with a programmable controller, push buttons, and an alphanumeric display. It is normally powered from the starter battery, but for the very highest class of installation a dedicated control battery and charger may be provided.

While the control of engine speed and generator voltage are undoubtedly part of the control system, these two functions are invariably performed directly by the speed governor and the voltage regulator. The speed governor and voltage regulator may interact with the control module for load and kVAr sharing, and for synchronizing purposes, which leaves the modular controller to perform other functions such as:

- Monitor the voltage and frequency of the normal supply
- Control the engine start and shut down sequences
- Open and close the switching devices controlling the normal and standby supplies, where appropriate initiate load shedding and apply the load to the generator in steps
- Monitor any warning alarms originating from the generating set and take appropriate action
- Monitor any shut down alarms originating from the generating set and take appropriate action
- Where a telemetry system is installed, transmit data to, and receive instructions from, the base station

The above functions are considered in detail in the following subsections.

Monitoring the Status of the Normal Supply

Monitoring the voltage and frequency of the normal supply should be at a point upstream of the normal supply changeover contactor or circuit breaker. If it is at a point downstream of the changeover device the restoration of the supply cannot be monitored. In addition the logic should be such as to prevent the generator attempting to supply the load in the event of the normal supply having tripped on fault or overload. This can be achieved by connecting an auxiliary contact on the protective device which passes information to the control module. It is usual to monitor the three phase voltages, any phase unbalance, and the frequency.

Automatic Starting and Shutdown Sequences

The automatic start sequence should commence after the expiry of an adjustable delay, (say 0–10 seconds), initiated by the normal supply mains monitor. The delay avoids spurious starting during a transient loss of the normal supply, and is particularly useful if the supply system includes any autoreclosing switchgear.

The automatic shutdown sequence is initiated either by the control system or by one or more of the engine safety shutdown devices. Some engines (particularly the larger units) are arranged to continue running off load for 5 or 10 min in order to allow heat to be removed from the cylinder block, heads, and pistons before finally shutting down. This procedure is not applicable to alarm shutdown signals such as overspeed, but the omission of the light load running period can lead to boiling of the cooling water at engine hot spots.

Operating the Switching Devices

After a failure of the normal supply the generating set should be automatically started, and the switching sequence to load the set completed, after which the standby supply remains in use until a shutdown command is given.

The switching sequence may involve three or more functional devices, which may be circuit breakers or contactors. They are the supply changeover devices, the load-shedding device(s), and any step load limiting devices necessary to apply the load to the set in two (or more) steps or stages. It is important that any load-shedding and step load limiting devices are open before the standby supply changeover device is closed.

If the switching devices are circuit breakers, the control system should arrange for their operation to be delayed until the generating set is up to speed and ready to accept load. This avoids unnecessary circuit breaker operation—if the normal supply is restored before the standby supply is available the generating set may be shut down and no circuit breaker operations are necessary.

For the shutdown sequence the first operation would be to open the standby supply switching device, thus unloading the set, followed by the closing of any load-shedding devices and, after a short delay, the closing of the normal supply switching device. The delay between opening one supply and closing the other is important. The two supplies are not synchronized and if one is replaced by the other without a break several problems can arise. Typical problems are:

- If the load includes any large induction motors they will be reconnected to the supply with their rotor fluxes in random phase relationship with the new supply. If the rotor flux is far removed from its synchronized position very large currents will flow and protective devices may operate. The rotor flux of an induction motor decays rapidly and the problem is avoided if there is a break of say 0.5 second between the two supplies.

- Any transformers or other static electromagnetic machines will experience a changeover of supplies in random phase relationship at the time of changeover. This can result in the magnetic circuits of the machines experiencing two successive half cycles of the same polarity leading to magnetic saturation, an excessive magnetizing current, and the operation of protective devices. The problem is avoided if there is a break of say 0.5 second between the two supplies.

- Any synchronous machines will experience a changeover of supplies in random phase relationship at the time of changeover. If the two supplies are far removed from synchronism very large currents will flow and protective devices may operate. To overcome this problem it may be necessary to allow such machines to run down to a stationary condition before connecting the new supply, and to restart it. This problem can theoretically arise when changing from normal to standby supply as well as from standby to normal supply. If a large synchronous motor is running light its inertia could very easily keep it running during the 10 to 15 seconds required for the standby supply to become available. It follows that any synchronous motor should be disconnected from its supply before any changeover of its supply is initiated.

Engine Warning and Alarm Shutdown Sequences

To protect the engine against catastrophic damage and to avoid unnecessary running when disconnected from its load the installation should be provided with devices which monitor the following parameters and pass appropriate signals to the control module:

- Low lubricating oil pressure
- High cooling water outlet temperature or, for air-cooled sets, high cylinder head temperature
- Operation of the engine overspeed trip
- Failure of the governor speed sensing signal magnetic pick-up
- Operation of the engine room fire detection system
- Tripping of the standby supply circuit breaker on fault or overload
- Failure to start after the appropriate cranking sequence

The first two parameters, low oil pressure and high coolant temperature may be monitored in two stages, the first stage being a warning, the second stage being a shut down. The warning stage is useful only if skilled personnel are available who can diagnose the condition and take remedial action.

Remote Control and Monitoring

Manufacturers of engines and generating sets are able to offer comprehensive control and monitoring systems operating from a central location. For continuously running plant such an arrangement can be very useful but for standby generating plant the advantages are obviously fewer.

Remote Control Systems and Test Runs

Depending on the degree of technical knowledge and ability that is available locally, it well may be worth considering the benefits of a data link to a remote base. The remote base can be staffed by the user organization, the plant manufacturer, or a specialist organization.

At locations where competent technical staff are available, the local staff can arrange test runs at regular intervals and in doing so they benefit from becoming familiar with the installation. A remote control system would be of little benefit. At locations where there is no on-site technical staff, a remote control system can initiate and complete regular test runs without the need for any local personnel becoming involved (except for giving permission for the test run). The system would involve monitoring the status of the installation to ensure that the set is ready to start; has adequate fuel, lubricating oil, and coolant; and that the electrical distribution system is in its normal healthy condition.

At locations which do not have technical staff and are isolated or difficult of access, a remote control facility could be useful. The staff responsible for operation and maintenance will be remote from the installation and will occasionally suffer from an overload of work. At such times it will be difficult to attach importance to a test run at a distant location, and it is likely that test runs will occasionally be neglected. With remote control the performance of test runs is so much simpler and the awkward wrangles are less likely to arise.

At locations which do not have technical staff an occasional visit by competent personnel will be necessary for inspection and maintenance purposes, but these do not have to coincide with test runs, they can be at any convenient time and frequency.

Remote Monitoring

In addition to remote control it is now possible to install remote monitoring systems. This is more appropriate to continuously running plant but a large standby generating set without the benefit of supervision from on-site technical staff may well deserve a degree of monitoring. Basic monitoring may include data such as starter battery condition, start and stop times, power, voltage, speed, exhaust temperature for each cylinder, jacket water temperature, oil pressure, engine room ambient temperature, and vibration.

If any of the readings do not fit into the expected pattern a warning is given to the appropriate person or persons. It is not necessary for monitoring to be continuous, data can be scanned periodically, say at 10-min intervals. Recorded data can be used to reconstruct events after a failure or breakdown.

Experienced personnel in the maintenance business tell us that veteran engine attendants are able to sense potential problems through unusual noise, vibration, or smell before the instrumentation has had time to react. We are in danger of losing this sixth sense!

Today's engines are likely to include microprocessor controls in the speed governing, voltage regulating, and fuel systems, and it is important that all control or monitoring systems within the installation comply with national legislation regarding electromagnetic compatibility and interference levels.

Location of Equipment

The location for the generating set should be carefully chosen. Where the standby installation is part of a new building or installation it will probably be housed in a generator room forming part of the new works. Where the standby installation is being added to an existing installation it is common practice to house the generating set in an outdoor container-type housing. The preceding sections of this book refer to a number of topics which have a bearing on the location of the equipment and are discussed in the text which follows.

Discharge of Exhaust Gases

The exhaust gases include a number of unpleasant pollutants and the discharge should be at high level, above the top of the building or housing and well away from any air inlets associated with building ventilation. In the United Kingdom the discharge should be in accordance with the 1956 Clean Air Act which was discussed in a previous section.

Noise from Exhaust Discharge

The exhaust outlet will be a source of noise over the full range of audible frequencies, dominant frequencies being the engine firing rate and its harmonics; a turbocharger will introduce higher frequencies. For a 1500-rpm six-cylinder four-stroke engine, the firing rate will be at 75 Hz, so it is apparent that the noise from an engine exhaust will include a wide spectrum of frequencies. The local authority will require information on the level of noise expected to be emitted.

In order to evaluate the effect of the noise on the neighborhood, it is necessary to obtain acoustic data relating to the emitted noise from the exhaust system design organization, which is usually the generating set manufacturer. The data should be in the form of acoustic sound pressure levels at a specified distance from the outlet, for octave band centre frequencies between 31.5 Hz and 8 kHz. The distance from the outlet will probably be 2 m and it may be useful to know that in a free field the noise is attenuated by 6 dB for each doubling of the distance. Thus at a distance of 4 m the attenuation would be 6 dB and at 16 m the attenuation would be 18 dB. The local authority will require information on the level of noise expected to be emitted and will have rules regarding noise levels which will take into account the type of nearby buildings and their occupants.

Noise from Diesel Engine Cooling and Ventilation Air

The radiator of a diesel engine will require a large quantity of cooling air. For sets up to about 700 kW the radiator will probably be local to the engine and the air passing through it will also provide ventilation for the engine room. If the radiator is remote from the engine it will have its own fan or fans which will be another source of noise and a supply of air will be required to the engine room to remove the heat gained from the engine casing. All these air inlets and outlets require openings in the enclosures which allow the escape of noise. Acoustic treatment will usually be required and the effect on the occupants of any nearby buildings will have to be considered. As in the case of exhaust noise the local authority will require information on the level of noise expected to be emitted.

Pollution from Diesel Engine Cooling and Ventilation Air

The air outlets will be a source of warm air polluted with various leakages and fumes from the hot engine surfaces; the outlets should be situated well away from any air inlets associated with building ventilation.

Gas Turbine Combustion and Ventilation Air

For gas turbines a similar approach should be taken, the amount of exhaust gas will be greater, there will be no radiator, but there will be a combustion air inlet and a ventilation air inlet and outlet, and air may be required for an oil cooler.

Vibration

Diesel engines, and to a lesser extent gas turbines, will produce vibration which to some degree will be transmitted to the building structure. Some thought should be given to this if the building is occupied or is in some other way sensitive to vibration.

Accessibility

The proposed site for the generating set must obviously allow for the initial installation of the set and should allow for future maintenance. Sometimes building work continues after installation and, if the work is of a structural nature, it can prevent the removal of a set in the unlikely event of a catastrophic failure. Installations exist where the engines are literally built into the building and cannot be removed, multiset installations exist where access is at one end of the room and the removal of the "first installed" set would involve the removal of the other sets. The avoidance of such installations demands an awareness of the problem at the early planning stage; these difficulties arise more often in congested, built-up areas such as city centers. The problem can sometimes be simplified by partly dismantling the set, the generator, the radiator, and even the cylinder heads can be removed fairly easily.

Other Considerations

Except for the smallest sets access to a supply of fuel will be needed; provided that a route exists this can be piped fairly easily from the bulk tanks. To avoid long distribution cables and associated voltage drops the set should be close to its electrical load.

Bibliography

International Standards

ISO 2314, BS 3135—Specification for gas turbine acceptance tests
ISO 3046, BS 5514—Reciprocating internal combustion engines. Performance.
Part 1 Standard reference conditions, declarations of power, fuel and lubricating oil consumption and test methods
Part 2 Not used
Part 3 Specification for test measurements
Part 4 Speed governing
Part 5 Torsional vibrations
Part 6 Specification for overspeed protection
Part 7 Codes for engine power
ISO 3977, BS 3863—Guide for gas turbine procurement
ISO 8528, BS 7698—Reciprocating internal combustion engine driven alternating current generating sets
Part 1 Specification for application, rating and performance
Part 2 Specification for engines
Part 3 Specification for alternating current generators for generating sets
Part 4 Specification for controlgear and switchgear
Part 5 Specification for generating sets
Part 6 Test methods
Part 7 Technical declarations for specification and design
Part 8 Requirements for tests for low power generating sets (up to 10 kW)
Part 9 Measurement and evaluation of mechanical vibration
Part 10 Measurement of airborne noise by the enveloping surface method
Part 11 Not used
Part 12 Emergency power supply to safety devices

British and European Standards

BS 2869 Specification for fuel oil for agricultural, domestic and industrial engines and boilers
BS EN 60034—Rotating electrical machines
Part 1—Rating and performance
Part 4—Methods for determining synchronous machine quantities from tests
Part 16—Excitation systems for synchronous machines
Part 22—Generators to be driven by reciprocating internal combustion engines

Electrical Generating Systems Association Standards

The following performance standards are published by the Electrical Generating Systems Association of Boca Raton, Florida

100B	Engine cranking batteries used with engine generator sets
100C	Battery chargers for engine starting batteries
100D	Generator overcurrent protection
100E	Governors on engines
100F	Engine protection schemes
100G	Generator set instrumentation, controls and auxiliary equipment
100M	Multiple engine generator set control systems
100P	Peak shaving controls
100R	Voltage regulators used on electric generators
100S	Transfer switches used with engine generator sets
100T	Diesel fuel systems for engine generator sets with above ground steel tanks

101P   Engine driven generator sets
101T   Generator set test methods

Other Documents

1956 Clean Air Act Memorandum (Third Edition). A U.K. document published by Her Majesty's Stationery Office.
Baranescu, R., and B. Challen, *Diesel Engine Reference Book*, 2nd ed. Butterworth-Heinemann, Oxford, United Kingdom, 1999.

2

Interconnecting the Standby and Normal Supplies

Introduction

This chapter is concerned with a number of matters that have to be considered before an installation can be designed. It includes such topics as the distribution of power to the essential and nonessential loads, the number of generating sets to be installed, their modes of operation and method of connection, the generated voltage, running in parallel with the normal supply, overcurrent protection, and the special characteristics of various loads.

Separating the Essential and Nonessential Loads

Before planning can proceed it is necessary to identify the essential loads (which receive a supply from the standby generator) and nonessential loads which are disconnected, and to devise a distribution system which will achieve the desired operation. Problems can arise and there are various methods of overcoming them.

If the standby supply is to be included in a new installation the simplest procedure is for the main switchboard to have two busbar sections, one supplied from the normal supply and feeding the nonessential load, the other supplied from the standby supply and feeding the essential load. The two sections of busbar are connected through a bus section circuit breaker which is electrically interlocked with the standby supply circuit breaker and acts as the load shedding device when the standby supply is in use. When it is necessary to run on the standby supply the set is started and when it is ready to accept load, the bus section breaker is tripped and, after a short delay, the standby supply breaker is closed. Figure 2.2 indicates such a system; for small installations interlocked changeover contactors may be used in the manner indicated in Fig. 2.1.

For a large installation where there are several major distribution boards, it may be expedient to equip other boards with busbars and circuit breakers as described above, and to provide a feed from the standby supply. This would involve the addition of a standby supply distribution board.

If a number of individual loads distributed within a building are classified as essential and have to be supplied from the standby supply, the simplest procedure is to install a supply changeover contactor for each of the loads. It is usually convenient to install the contactors near to their loads but such a decision depends on where the normal and standby supplies are available. The contactor control circuits may be autonomous or may receive signals from the generator control panel. If they are autonomous they should be biased to use the normal supply whenever it is available, the standby supply being used only when the normal supply is not available. For autonomous controls there should be two timers, one to delay the closing of the standby supply contactor so that the loads may be applied to the generating set in stages, and another to delay the return to the normal supply for sufficient time to prove its reliability.

Another consideration that sometimes arises is the geographical nature of the site. If the installation is "compact," such as a single building, the foregoing considerations will be applicable but if it is, for instance, an airfield, it is "spread out" and other considerations arise. In addition to using split bus distribution boards, a large, spread-out site may use a signal distributed over the site by, for example, a telephone-type cable network, the final shedding of the nonessential load being achieved by using various devices such as:

- *Circuit breakers with shunt trips.* These are applicable to both low-voltage and high-voltage installations and will usually incorporate a facility enabling them to be closed automatically or manually from a remote control point.

- *Ring main units with shunt trips.* These are applicable to high-voltage installations only. A remote closing facility may not be available on ring main units and in such cases reclosing involves an operator visiting the location of each ring main unit.

- *Outdoor-type fuse pillars having duplicate busbars* (the nonessential bars being supplied through a contactor). These are applicable to low-voltage circuits only. The contactor coil is supplied through a normally closed contact of a load shedding relay. A load shedding signal energizes the relay thus releasing the contactor which remains open until the load shedding signal is removed. A circuit breaker may be used in place of the contactor. For a large site using many load shedding relays and long runs of telephone type cable, consideration must be given to the relay operating current and the volt drop in the telephone-type cable.

Figures 2.1 and 2.2 show typical methods of connecting single generating sets, Fig. 2.1 applies to a small set using electrically and mechanically interlocked changeover contactors or circuit breakers, and Fig. 2.2 applies to a set using electrically interlocked circuit breakers.

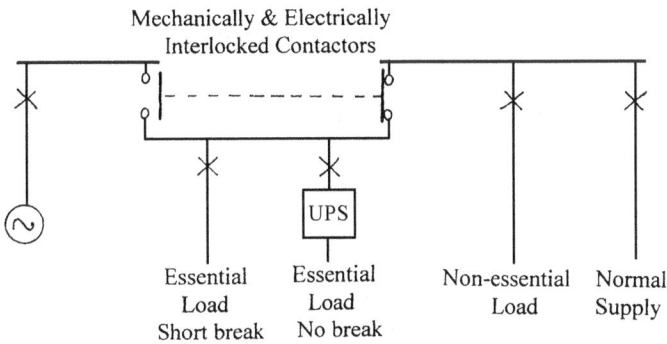

Figure 2.1   Single-line diagram showing a typical method of connecting a small single set using changeover contactors.

Circuit Breaker	CB1	CB2	CB3	CB4	CB5
Normal	O	I	1	1	1
Standby	I	I	O	O or I	O or I

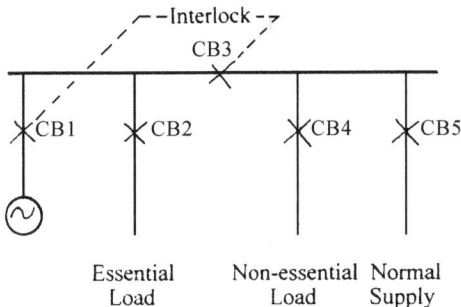

Figure 2.2   Single-line diagram and switching status table showing a typical method of connecting a single set using electrically interlocked circuit breakers.

Circuit Breaker	CB1	CB2	CB3	CB4	CB5	CB6	CB7	CB8
Normal	O	O	O	I	I	I	I	I
Standby 2 Sets	I	I	I	O	O or 1	I	I	O or I
Standby 1 Set	I	O	I	O	O or 1	I	O	O or I

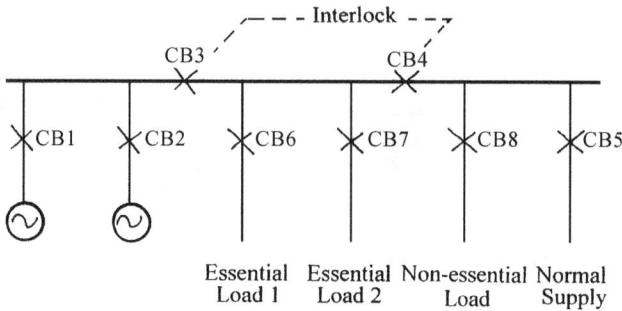

Figure 2.3 Single-line diagram and switching status table showing a typical method of connecting two sets with two stages of load shedding.

Where two generating sets are installed and run in parallel, two stages of load shedding may be provided; the essential load is divided into two parts which may be termed essential load 1 and essential load 2. On failure of the normal supply, the nonessential load is shed and essential loads 1 and 2 are fed from the standby supply, and if either of the sets fails essential load 2 is shed leaving essential load 1 to be supplied from the healthy set. Figure 2.3 indicates a typical electrical arrangement.

Use of Multiple Generating Sets

Reasons for Using Multiple Sets

Most standby power installations use a single generating set; this is the simplest arrangement. A single set provides no redundancy and its failure deprives the entire installation of its standby supply; two or more sets are sometimes used for one or more of the following reasons:

- A single set provides no redundancy and its failure deprives the entire installation of its standby supply. Very important installations may use two sets, each rated for the essential load, this provides 100 percent redundancy. On loss of supply both sets are started and the first set available is connected to the essential load; after the first set has accepted the load the second set is shut down. Other arrangements may be used, for example three sets each rated for one half of the essential load, which provides 50 percent redundancy.

- The use of two sets, each rated for one half of the essential load, allows part of the installation to be supplied with standby power after the failure of one set. This assumes that the installation is such that it can continue to operate on a reduced scale. On loss of supply both sets are started and either paralleled before the essential load is applied, or the first available set supplies essential load 1, the second set is then paralleled and essential load 2 is connected.

- If a diesel engine is run on light load for a sustained period of time, products of combustion are deposited on the cylinder bores and lead to "glazing" and lubrication problems. In addition combustion may be incomplete, leading to carbon deposits within the combustion chambers. Where the electrical load is likely to vary and the lower limit approaches one third of the upper limit, the use of two sets should be considered, each rated at say 50 percent of the essential loading. On light load one set can then be shut down, but note that the second set has to be restored, either manually or automatically, before the full essential load can be supplied. If redundancy is required three sets would be used.

- The use of two independently running sets makes it possible to separate industrial-type loads from loads requiring a clean supply. The two sets may be of different sizes and to allow for the failure of either set, manual switching can be provided to rearrange the loading on the healthy set.

The use of two sets doubles the probability of a failure of one set but the probability of a total failure of the standby supply is very much reduced.

Modes of Operation

Where multiple sets are required to be paralleled before supplying the essential load, time will be required for the paralleling procedure and the time between the loss of normal supply and the availability of the standby supply will be greater than that which occurs for a single set of similar rating. The maximum time for paralleling one set to the busbars will probably be on the order of a minute. Most critical loads nowadays are supplied through an uninterruptible power supply with a long period of autonomy, and battery-maintained emergency lighting is normal in buildings. In such installations the duration of the "dead" period cannot be claimed to be important. It probably makes little difference whether the break in supply is 15 seconds, 30 seconds, or a minute.

If, however, in a two set installation it is required to keep the "dead" period to a minimum, an alternative mode of loading can be used. On loss of supply both sets are started and the first available set supplies essential load 1, the second set is then paralleled and essential load 2 is connected.

Load Sharing

When more than one set supplies the load, they will usually be of the same rating in which case they should share the kilowatt and kilovar loads equally. If they are of different ratings they should share the loads in proportion to their ratings.

The simplest system of kilowatt load sharing is for the engine speed governors to allow a small drop in speed as the load increases as indicated by Fig. 2.4. The paralleled sets must of necessity run at the same speed and if each set has a similar load/speed characteristic, load sharing will automatically be achieved. If at some time the load sharing between two sets becomes unequal, both sets will be running at speeds which are not compatible with their load/speed characteristics and their governors will react to correct the errors.

If the paralleled sets are of the different ratings, provided that the droop settings of each of the sets are the same, the load will be shared proportionally to their ratings. It can be seen from Fig. 2.4 that the load taken by any set can be increased or decreased by reducing or increasing the droop setting (the slope of the characteristic) or by altering the set speed (the vertical position of the characteristic).

If the frequency droop is not acceptable, an alternative system of sharing the kilowatt loading is for the engine speed governors to be arranged for isochronous (constant speed) running. This requires electrical interconnections between the governors to maintain load sharing.

Kilovar load sharing between generators running in parallel is dependent upon the rotor excitation which in turn is determined by the voltage regulators. As with kW load sharing the simplest system is for the

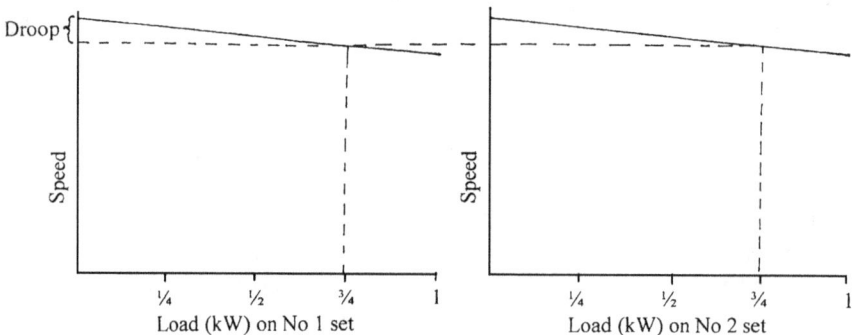

Figure 2.4 Kilowatt load sharing between two sets using a speed droop governor characteristic.

voltage regulators to introduce a small drop in terminal voltage as the kilovar load increases as indicated by Fig. 2.5. The difference between Figs. 2.4 and 2.5 is that in the former the axes represent speed and kW loading, and in the latter they represent voltage and kVAr loading. As the paralleled sets will be running at the same voltage, if they have the same droop settings, load sharing will be achieved automatically. One method of achieving the droop is for each regulator to be provided with a signal from a quadrature drop current transformer connected in one phase of its generator output. The quadrature drop signal is added to the terminal voltage sampled by the voltage regulator and thereby introduces a voltage droop.

If the voltage droop is not acceptable, an alternative system of sharing the kilovar loading is used, this requires interconnections between the voltage regulators and maintains constant voltage. If one machine is taking too great a share of kilovars the sharing signal will cause its regulator to reduce its excitation.

Where frequency and voltage droop are used to share load and kVAr the droop will be, say, 3 or 4 percent and it is usual to arrange for the nominal frequency and voltage to be achieved at normal running load.

Interconnections with the Normal Supply

The main power interconnections between the normal supply, the standby supply, and the essential and nonessential loads depend on the number of sets in use and whether the power changeover is performed with two circuit breakers or an arrangement of contactors. Where the normal and standby supplies are required to be paralleled, the following comments are applicable but the reader should in addition refer to the section titled "Paralleling the Standby and Normal Supplies."

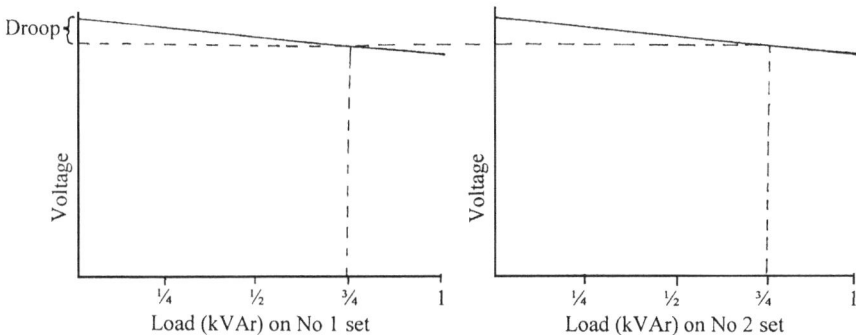

Figure 2.5   Kilovar load sharing between two sets using a voltage regulator with a quadrature droop characteristic.

The changeover devices may require three or four poles depending upon the type of supply and the details of the agreement between the Embedded Generator and the network operator. The matter is discussed in the section titled "Neutral Connections for Single Sets Not Intended to Run in Parallel with the Normal Supply."

Use of Changeover Contactors

Changeover contactors are suitable for use with small generating sets having a current rating not more than a few hundred amperes. Figure 2.1 indicates a typical interconnecting arrangement.

If the contacts of a contactor suffer misuse or severe arcing, welding can result; for this reason contactors switching the normal and standby supplies must be electrically and mechanically interlocked. The use of mechanically interlocked changeover contactors requires that both the supplies (normal and standby) are taken to the same component which therefore becomes critical. Any failure of the contactor results in neither supply being available and a total shutdown is required for repair (or indeed for maintenance).

Use of Electrically Interlocked Circuit Breakers

The use of circuit breakers is applicable to generating sets of any rating; the contact arrangement and the operating mechanism are more robust than that of contactors and mechanical interlocking is not essential. The normal and standby supplies are taken to different components and the failure of one circuit breaker does not necessarily render the other inoperative.

Molded-case circuit breakers may be mechanically and electrically interlocked but this arrangement results in the two circuit breakers being mounted close to each other and, unless live working is allowed, a shut down will be required for repair or maintenance. For mechanically interlocked circuit breakers the arrangement is similar to that used for contactors and Fig. 2.1 applies.

Air circuit breakers constructed within a metal frame and intended for switchboard mounting cannot usually be mechanically interlocked, and reliance must be on electrical interlocking. Where there is no mechanical interlocking the two circuit breakers are independent and do not have to be mounted together. Figures 2.2 and 2.3 indicate typical interconnecting arrangements.

In Figs. 2.2 and 2.3 the normal supply is taken to the busbar section that feeds the nonessential load, which results in an operating advantage; while the essential load is being supplied from the standby supply, on restoration of the normal supply it is now permissible and possible to close the normal supply circuit breaker and supply the nonessential load. Indeed, it is not necessary to trip the normal supply circuit breaker because the two supplies are separated by bus section breaker CB4, on restoration of the normal supply the nonessential load is then energized.

The Fig. 2.3 interconnections provide another operating advantage. The generator section of the busbars does not have any other circuits connected to it; this allows the two generating sets to be run up to speed, paralleled and connected to the busbars for test purposes without interfering with normal running.

A key-operated manual override facility can be added to electrically interlocked circuit breakers and some uses for such a system are described in the following paragraphs. They require the presence of skilled personnel. Each key must be unique, not interchangeable with others used locally and must be clearly and permanently identifiable. Where the override involves a "two out of three" selection a separately mounted key exchange mechanism is provided; this holds the three keys but allows only two to be extracted.

Alternative Operating Regimes

Many variations of the interconnections with the normal supply are possible. In Fig. 2.2 changing the electrical interlock from CB1 and CB3 to CB1 and CB5 makes it possible, during standby running, to use surplus standby power to feed selected items of the nonessential load. This is achieved at the expense of losing the ability to restore the normal supply to the nonessential load before shutting down the standby supply. Similar reasoning can be applied to Fig. 2.3.

A more complex system can be devised which retains the ability to use surplus standby power to feed selected items of the nonessential load or to restore the normal power to the nonessential load before shutting down the standby supply. Referring to Fig. 2.2, if the electrical interlock between CB1 and CB3 is retained for automatic operation, but a manual override is provided which replaces it with a key interlock which allows two, and only two, circuit breakers of the set CB1, CB3, and CB5 to be closed at any time, versatile operation is possible. Considering only these three circuit breakers the following rules apply:

In normal operation CB3 and CB5 are closed

In standby operation only CB1 is closed

To supply some nonessential load from the standby supply CB1 and CB3 are closed

To use the normal and standby supplies at the same time CB1 and CB5 are closed.

Operating regimes such as those described in the last two paragraphs should not be used indiscriminately, they require skilled and competent operating personnel who have the necessary experience and confidence to operate switchgear under the somewhat stressful conditions that always apply after a failure of the normal supply. For the benefit of operating personnel a wall chart should display relevant information such as that given above.

The Generator Voltage

Most standby generating sets are of a rating well below 1 MW and will be connected to systems using the standard final distribution voltage which, in the United Kingdom, is 400/230 volts. However generating sets having a rating approaching 1 MW may be associated with a high-voltage distribution system. In the United Kingdom, the predominant high-voltage distribution voltage is 11 kV, but 20kV is used in some rural areas.

Each generator manufacturer will have determined a rating at which the transition from low voltage to high voltage occurs on grounds of costs and practicability. A generator designed for say, 100 kW at high voltage would comprise much insulation and very little copper; likewise a generator designed for say 5 MVA at low voltage would comprise much copper and very little insulation. For every rating there is an optimum voltage but choice is usually limited to the standard distribution voltages which in the United Kingdom are 400 V, 11 kV, and 20 kV. The precise rating at which the transition occurs depends on details of manufacturing techniques but it is likely to be of the order of 1250 kVA.

The current rating of a 1250-kVA generator at 400 V is 1804 amperes and cables will be of large cross section and intractable and may be difficult to install. If during the planning of an installation such difficulties are foreseen, it may be worth considering generating at low voltage and feeding directly into a step-up generator transformer. However, unless high voltage is already in use within the installation, its introduction is likely to cause problems with the competence of personnel and operating and safety procedures.

The Electricity Supply Regulations 1988

The Electricity Supply Regulations are currently applicable, within the United Kingdom, to suppliers of electricity, but they are soon to be replaced by the Electricity Safety, Quality and Continuity Regulations. These notes are included here as they indicate the underlying philosophy which seems unlikely to be changed by the new document.

The Regulations set out the requirements for the distribution of electricity, from its origin (a generating station or a substation) to the various consumers' installations, in order to ensure the safety of the supplier, the consumers, and any other personnel who may be involved. The following text is not intended as a summary of the Regulations; the text draws attention to those parts likely to be of particular interest to installers of standby generating plant.

Within the Regulations high voltage is defined as being above 1000 V ac.

Earthing

It is required that both high-voltage and low-voltage supplies shall be connected with earth at, or as near as is reasonably practicable to, the source of voltage. When high- and low-voltage systems are to be connected with earth in the same location there are two requirements:

1. The two earth systems shall not be interconnected unless the combined earth resistance does not exceed 1 ohm.

2. The systems shall not be connected to separate earth electrodes unless the overlap of the areas of resistance is not sufficient to cause danger.

Multiple Earthing

The supplier may connect the supply neutral with earth at places in addition to those described under the heading "Earthing" above, provided that the supply neutral conductor for three-phase circuits has a cross-sectional area not less than one-half of the phase conductors. For single-phase circuits the neutral shall have a cross-sectional area of not less than the phase conductor.

(*Note*: The minimum cross-sectional area requirement leads to a limitation of voltage difference between the earthing electrodes.)

There are special requirements for protective multiple earthing.

Protection against Overload, Earth Leakage, and Excess Voltage

"The public electricity supplier is to include protective devices in every system to prevent, so far as is reasonably practicable, danger arising from excess current or from current leakage to earth. Similarly, the supplier is to make arrangements which ensure, so far as is reasonably practicable, that danger will not arise from accidental contact between low- and high-voltage conductors.

The Consumer's Installation

The supplier is to be satisfied that the consumer's installation is properly installed, safe to use, and does not present danger. If an alternative supply is available within the consumer's installation, two forms of connection are recognized:

1. The alternative connection in which none of the conductors of one supply (except those connected with earth) are ever connected with the conductors of the other supply.

2. The parallel connection which allows the two supplies to be interconnected, provided that a suitable written agreement exists between the two suppliers, and that the maintenance and operation of the plant is undertaken in a competent and safe manner. The agreement is to include such matters as the means of synchronizing, earthing, maintenance records, competence of personnel, and means of communication between operators of the interconnected supplies."

Engineering Recomendation G.59/1

The full title of this document is "Recommendations for the connection of embedded generating plant to the public electricity suppliers' distribution systems." It originated in 1985 as Engineering Recommendation G.59, and following the privatization of the electricity supply industry, was revised in 1991 by the Electricity Association of the United Kingdom who now publish it as Engineering Recommendation G.59/1 (ER G.59/1). It is intended for the use of the U.K. public electricity suppliers and their consumers and applies to generating plants not exceeding 5 MW rating which is connected to systems operating at 20 kV or below.

The term "embedded generating plant" is used to describe any generating plant connected to a public electricity supply, whether it is intended for parallel operation or not. Persons who operate such a plant are defined as *embedded generators*. Three types of connection are recognized:

- The alternative connection in which the embedded generator operates as an alternative to the public electricity supply. The arrangement must be such that the two supplies cannot in any circumstances be paralleled.

- The parallel connection in which the embedded generator may run in parallel with the public electricity supply for unlimited periods. For this mode of operation an important consideration is the safety of the public electricity supplier's personnel who may find themselves working on a distribution system which unexpectedly becomes connected to the embedded generating plant.

- Occasional paralleling in which the embedded generator may run in parallel with the public electricity supply for a limited period, typically 5 min, and only for the purposes of maintaining continuity of supply while changing over from one source to the other.

For standby power installations interest is limited to the alternative connection and to occasional paralleling. The document describes the technical requirements for earthing the standby supply, interconnecting the neutrals of the two supplies and the changeover devices, and the electrical protection required during parallel running. These matters are discussed in the sections titled "Earthing the Neutral of the Standby Supply," "Neutral Connections for Single Sets Not Intended to Run in Parallel with the Normal Supply," and "Paralleling the Standby and Normal Supplies."

There is an associated guidance document Engineering Technical Report 113 (ETR 113) which discusses and describes these matters in detail. Many of the diagrams reproduced in this chapter have been derived from diagrams appearing in ETR 113.

Earthing the Neutral of the Standby Supply

It is important that the standby supply has a reliable earthing system; it is not safe to rely on any earthing arrangements that are not under the control of the site operating personnel. If the earthing system for the normal supply is local to the generating set and is under the control of the site operating personnel it may be permissible to use it as the earth for the standby supply. Figure 2.6 illustrates typical

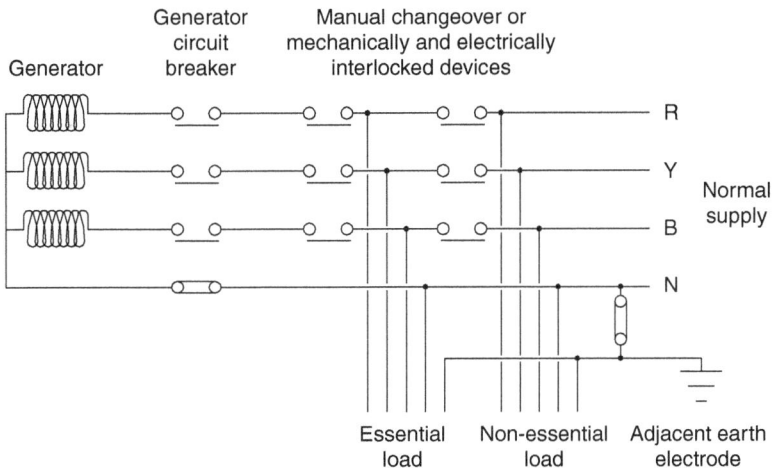

Figure 2.6 Typical connections for alternative supplies when the normal supply is from an adjacent substation with its own local earth electrode.

connections that can be used, the agreement of the network operator having been obtained. If the normal supply earthing conductor uses the sheath or armor of a supply cable that has a remote origin it cannot be relied upon. Such a cable may be disconnected by the network operator during maintenance or may be severed accidentally by a contractor (on or off site).

Earthing Low-Voltage Supplies

A low-voltage standby supply that runs independently of the normal supply should have its neutral solidly connected to a local earth electrode. If there are separate earthing systems associated with the normal supply and the standby supply they should be bonded together. When devising the interconnections between the two neutrals it must be kept in mind that the network operator will not allow its neutral to be connected to earth at a second point unless the system incorporates protective multiple earthing. If the system does not incorporate protective multiple earthing a four-pole changeover device will be required as described in the next section. Within the United Kingdom protective multiple earthing is normal and in such cases the standby supply earth electrode should have a resistance to earth not exceeding 20 ohms. This is the resistance allowed for protective multiple earthing in the Electricity Supply Regulations.

Where it is intended to run the standby and normal supplies in parallel there are additional requirements which are described in the section titled "Paralleling the Standby and Normal Supplies."

Earthing High-Voltage Supplies

A high-voltage standby supply that runs independently of the normal supply usually has its neutral earthed through an earthing resistor which limits any earth fault currents to rated current. The earthing resistor provides two benefits, in the event of an internal fault the damage to the machine is limited, and in the event of an external fault the rise of potential of exposed conductive parts is limited. It should be noted that the earthing conductor between the star point and the earthing resistor will, under fault conditions, experience a rise of potential, the magnitude will depend upon the position of the fault; the earthing conductor should therefore be insulated for the phase voltage.

Where it is intended to run the standby and normal supplies in parallel there are additional requirements which are described in the section titled "Paralleling the Standby and Normal Supplies."

Within an installation, the Electricity Supply Regulations allow the high-voltage earth electrode system to be interconnected with the low-voltage earth electrode system provided that the combined resistance to earth does not exceed 1 ohm. If the resistance exceeds 1 ohm the Regulations forbid any interconnection and require the two electrode systems to be separated sufficiently to ensure that the overlap between the two resistance areas does not cause danger. In most industrial installations it is not practicable to ensure that the two earthing systems are not interconnected, or that the electrode resistance areas do not overlap, in such cases it follows that a combined resistance not exceeding 1 ohm is required for high-voltage neutral earthing.

Neutral Conections for Single Sets Not Intended to Run in Parallel with the Normal Supply

Where the standby and the normal supplies are not arranged to run in parallel Engineering Recommendation G59/1 describes this as the alternative connection.

Low-Voltage Installations

The neutral of a low-voltage standby supply should be solidly earthed. In a single set low-voltage installation the generator will usually be connected, as a four-wire machine, to a distribution board with the star point connected to the neutral busbar which in turn is connected to the earth bar and on to the earth electrode (see Fig. 2.9a).

Within the United Kingdom four systems of public supply may be encountered, they are described in BS 7671 and are:

- TN-C-S systems in which the neutral and the earthed protective conductors are combined into a single conductor in part of the system. This is the most common system and uses protective multiple earthing (PME). Where a standby supply is installed in such a system a triple pole changeover device will be required to select one supply or the other as illustrated by Fig. 2.7.

- TN-C systems in which the neutral and the earthed protective conductors are combined throughout the system. Provided that the system uses multiple earthing the same considerations apply as to the preceding TN-C-S systems.

- TN-S systems in which the neutral and the earthed protective conductors remain separate throughout the system. Where a standby supply is installed in such a system a four-pole changeover device

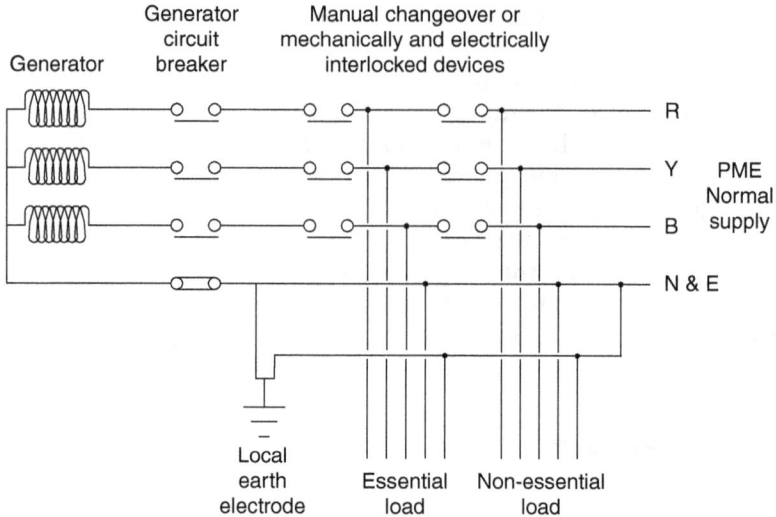

Figure 2.7 Typical connections for alternative supplies when the normal supply uses multiple earthing (PME, TN-C-S, or TN-C).

will be required to select one supply or the other as illustrated by Fig. 2.8. The fourth pole is required to avoid earthing the neutral at a second point.

- TT systems in which the neutral is earthed at the power source but the network operator does not provide an earthed protective conductor. Where a standby supply is installed in such a system a four-pole changeover device will be required (Fig. 2.8). The fourth pole is required to avoid earthing the neutral at a second point.

Figures 2.6, 2.7, and 2.8 have been derived from diagrams appearing in ETR113. It will be noted that Fig. 2.7 provides a parallel path for the neutral current, which may have undesirable consequences. One path uses the neutral conductor but there is another path which uses the protective conductors connecting to the local earth. The current in the protective conductors will be indeterminate, it may exceed the rating of the protective conductor and may upset the current balance in any protective residual current devices. The difficulty can be overcome by using four-pole changeover devices.

High-Voltage Installations

In a single set high-voltage installation the generator will usually be connected, as a three-wire machine, to a distribution board. The star

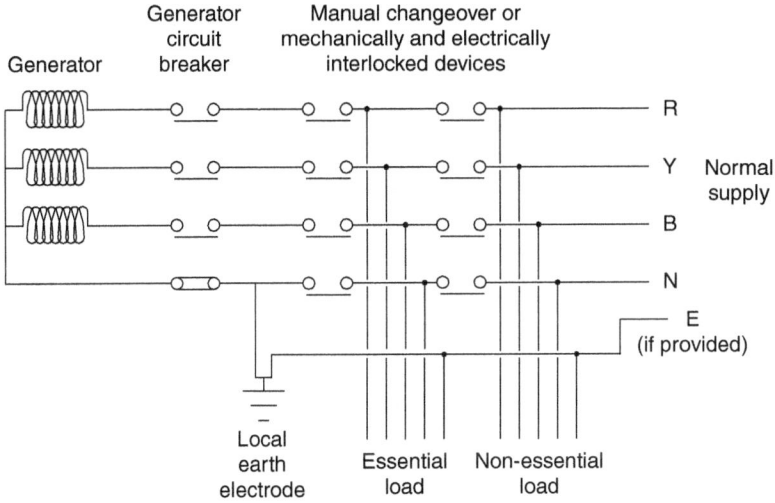

Figure 2.8 Typical connections for alternative supplies when the normal supply uses single point earthing (TN-S or TT).

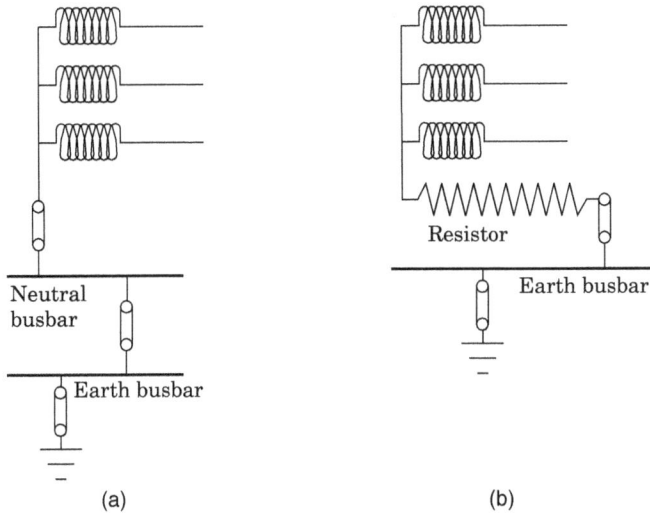

(a)   (b)

Figure 2.9 Neutral earthing of a generator. (*a*) Low voltage. (*b*) High voltage.

point will be connected to the earthing resistor which in turn will be connected to the earth bar and on to the earth electrode (see Figure 2.9*b*).

High-voltage power is usually supplied from a three-wire system, there being no neutral, an earth connection may not always be provided. For such installations triple pole electrically interlocked circuit breakers will be required as illustrated by Fig. 2.10.

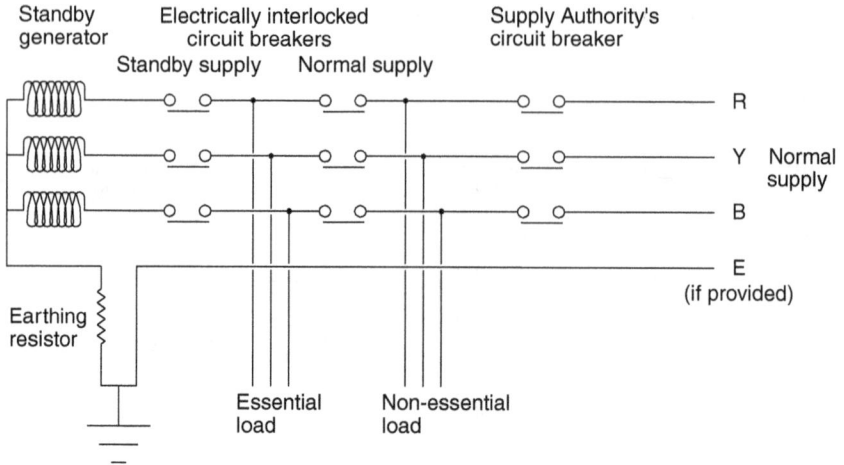

Figure 2.10 Typical connections for alternative supplies at high voltage.

Neutral Connections for Multiple Sets Not Intended to Run in Parallel with the Normal Supply

Running Generators in Parallel- and Triplen-Current Flow

Triplen currents have not been previously mentioned and it is appropriate to introduce them here. The harmonic orders 3, 6, 9, 12, etc. are known as triplen harmonics or triplens. They have zero-phase sequence and triplen currents flowing in the phase conductors become additive in the neutral, hence the concern expressed in the following text.

The voltage generated by a loaded salient pole generator will include a third harmonic and other triplen components the magnitude of which will be dependent upon the machine design, its excitation, and its loading. If the machines are identical and are equally loaded and excited the triplen harmonic voltages in each of the machines will be equal, in phase and balanced and there will be no resulting current flow. If the machines are not identical or are not equally loaded or excited, triplen harmonic currents will circulate between the machines as shown in Fig. 2.11. Any triplen load currents taken from them will similarly be additive in the neutral busbar return path as shown in Fig. 2.12.

It is likely that both the above conditions will exist at the same time and for the triplen currents in the neutral connections to be unexpectedly large. When salient pole generators are to be run in parallel, consideration should be given to the possibility of undesirable third harmonic currents circulating in the neutral connections of the machines. If the neutrals of paralleled generators are not

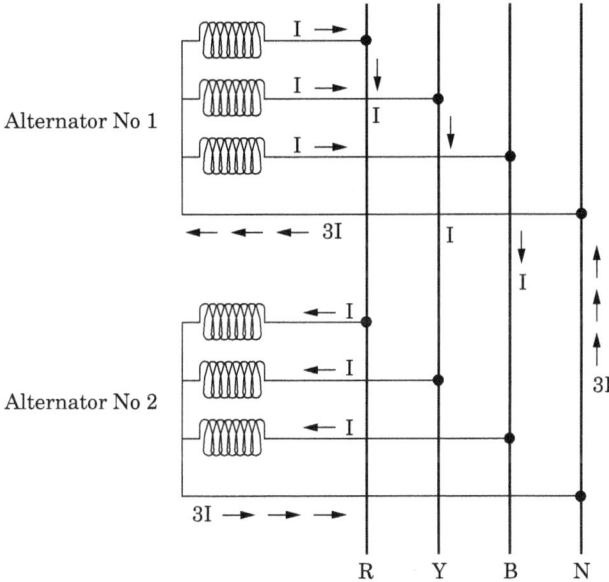

Figure 2.11 Triplen harmonic currents flowing in the windings of paralleled generators due to differing machine waveforms become additive in the neutral.

interconnected there can be no triplen harmonic circulating currents and the problem will not arise.

If the machines are identical and are running under identical excitation and loading conditions their neutral terminals may be safely connected together. However machines which are thought to be identical may differ and the excitation and loading conditions will be subject to the normal tolerances of the voltage regulators and load sharing systems. The failure of a voltage regulator, a speed governor, a load sharing system, or a kVAr sharing system may cause unexpectedly heavy circulating currents to flow in the neutral connections. For these reasons, the generator manufacturer should be consulted before the neutral terminals of paralleled generators are connected together.

Low-Voltage Installations

The supply should be solidly earthed as described for the single set installation but, for the reasons described in the preceding paragraphs, it may not be advisable to connect together the machine neutrals of a multiple set installation. There are several methods of overcoming the difficulty, the matter should be agreed with the generator manufacturer:

- Each machine is provided with a neutral contactor but only one contactor is closed at any time. A logic system ensures that the neutral

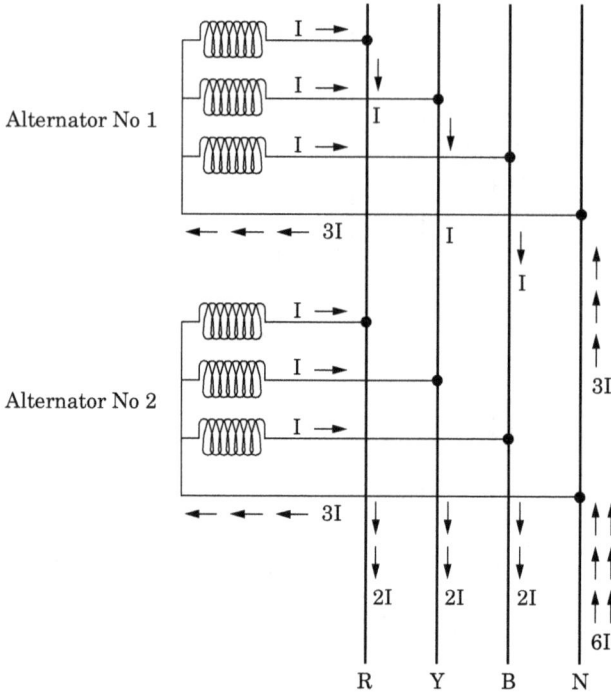

Figure 2.12 Triplen harmonic currents flowing in the load become additive in the neutral.

of one machine is connected and decides which machine it is; it will usually be the first generator on line. With this method the connected machine necessarily accepts all the neutral or out of balance current; if the neutral current is likely to exceed the neutral current rating of a single machine the method cannot be used. It should be noted that the neutral current to be considered is the rms sum of the fundamental neutral current due to unbalanced loading between the phases and the triplen neutral current generated by the load. Figure 2.13 illustrates the electrical connections.

- The supply is earthed through a static balancer having an interstar winding. The generator neutral terminals are not used and the star point of the static balancer becomes the neutral of the supply, and is solidly earthed. The static balancer divides any neutral load current into three equal zero sequence components which return to the load as part of the phase currents and do not pass through the generator windings. The generators share the positive and negative sequence components of the load current, the static balancer affecting only the zero sequence components. Figure 2.14 indicates the electrical connections.

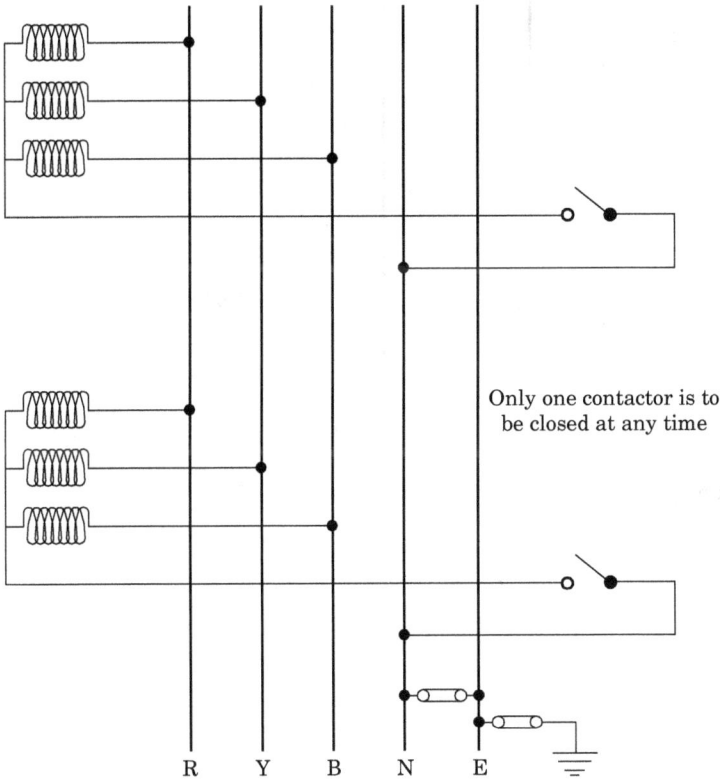

Figure 2.13 Connecting the neutral for low-voltage multiple sets using contactors.

The static balancer is a vital component connected to the main busbars and thought should be given to the electrical protection that should be provided. If it is tripped from the busbars due to a fault the standby supply has to be shut down due to the lack of a neutral connection. It follows that any protection such as overcurrent or earth fault should have current settings and time delays which avoid the probability of spurious tripping.

- Each machine is provided with a reactor connected between its star point and the neutral busbar. These will attenuate the third and higher harmonic currents without offering significant impedance at fundamental frequency. The reactors have the effect of increasing the generator's zero sequence impedance and data will normally be required from the generator manufacturer before any calculations can be undertaken. If there is fundamental current flow in the neutral due to load imbalance between the phases the reactors will increase the voltage unbalance—a compromise has to be reached between unbalanced phase voltages and triplen current flow.

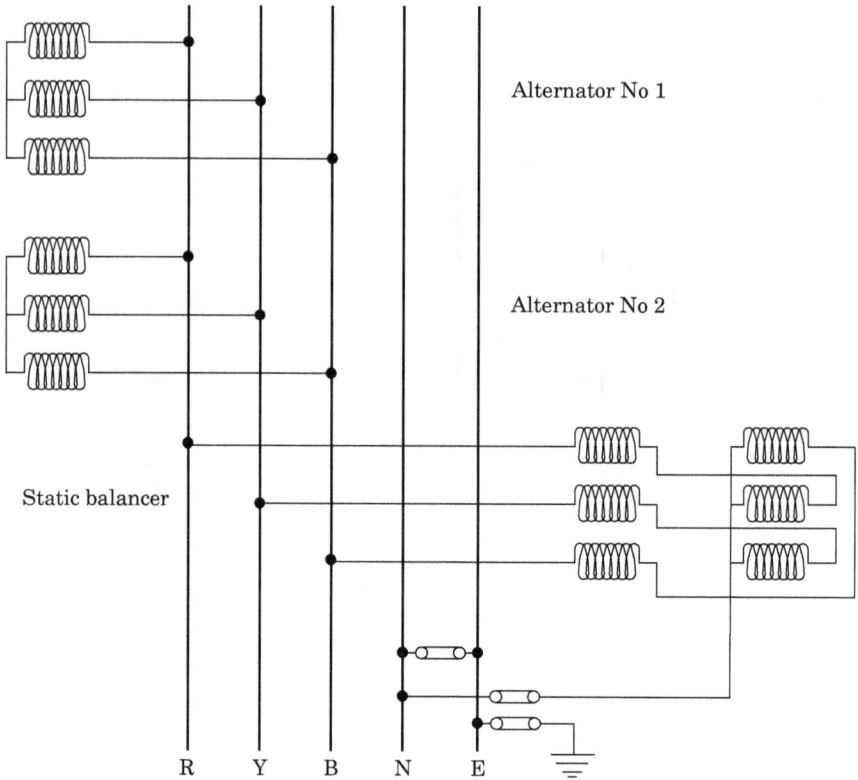

Figure 2.14   Connecting the neutral for low-voltage multiple sets using a static balancer.

- With the generator maker's approval the generator neutrals are paralleled. Unless the generators are identical and are working under identical conditions of excitation and loading there will be some third harmonic currents circulating in the neutrals.

High-Voltage Installations

The supply should be earthed through an earthing resistor as described for the single set installation but, for the reasons described in a foregoing paragraph, it may not be advisable to connect together the star points of a multiple set installation. There are several methods of overcoming the difficulty, the matter should be agreed with the generator manufacturer:

- Each machine is provided with an earthing contactor connected to a common earthing resistor but only one contactor is closed at any time. A logic system ensures that one machine is earthed and decides which machine it is; it will usually be the first generator on line. The resistor should be rated to limit any earth current to the current rating of one machine. Figure 2.15 illustrates the electrical connections.

- Each generator is earthed through its own earthing resistor. Circulating currents will be limited to an acceptable level by the resistors and the complications of contactors and a logic system are avoided. The circulating currents are unlikely to be high but the rating of the resistors should be continuously rated to dissipate the resulting losses. Figure 2.16 illustrates the electrical connections.

- The supply is earthed through an earthing transformer, which is similar to the low-voltage static balancer but is not rated to carry any neutral current. The generator star point terminals are not used and the star point of the earthing transformer is connected to the earthing resistor.

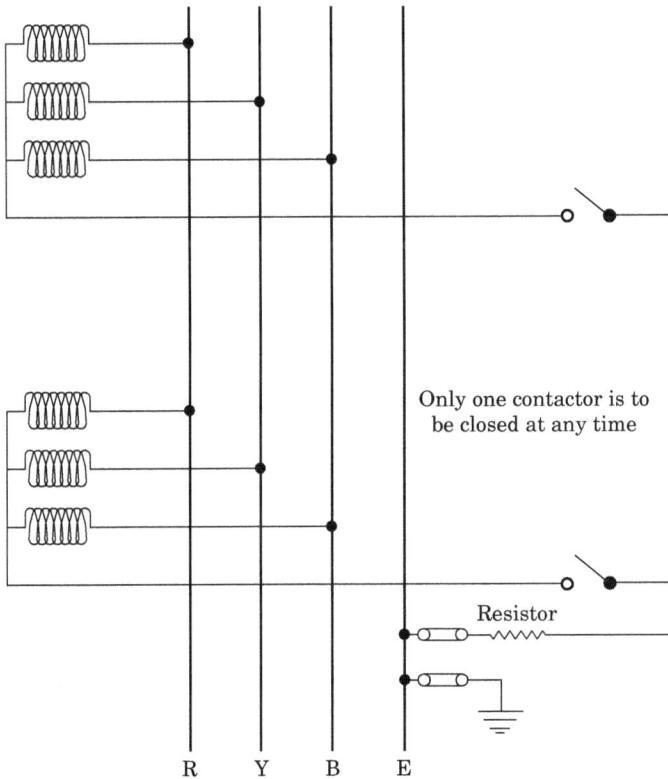

Figure 2.15 Neutral earthing of high-voltage multiple sets using contactors and a common resistor.

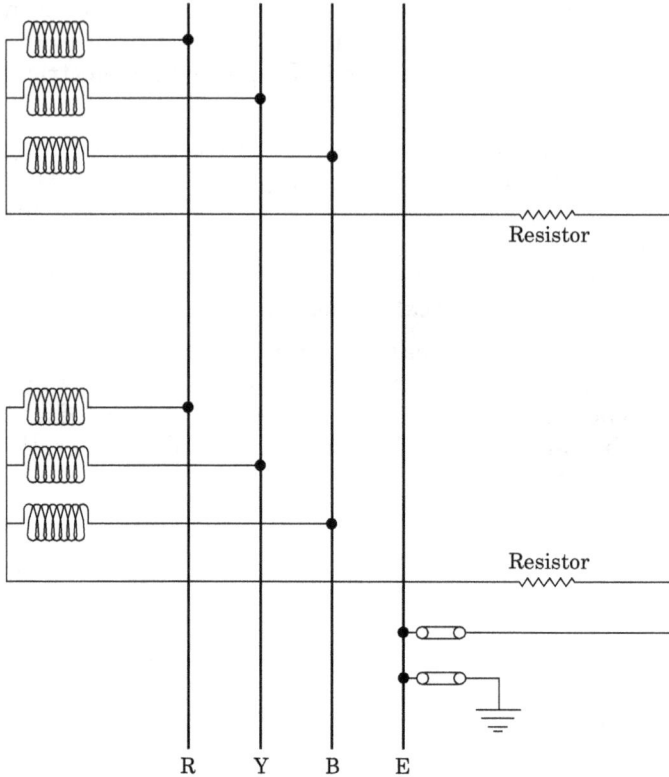

Figure 2.16 Neutral earthing of high-voltage multiple sets each generator having its own resistor.

Paralleling the Standby and Normal Supplies

Introduction

A facility that allows the standby supply to be occasionally run in parallel with the normal supply is extremely useful. Although paralleling is applicable to installations of any rating, in practice the facility is used with larger installations having ratings of say 500 kVA and above. It will require the agreement of the network operator and there will be requirements regarding the earthing arrangements, the protection relays fitted to the interconnecting circuit breaker and the operating procedures. The network operator must be satisfied that the operating personnel will be competent and will be available as and when necessary.

The advantage to be gained from the paralleling facility is the ability to transfer the load from one supply to the other without the load experiencing a break in supply. After running on the standby supply it is possible to return the essential load to the normal supply without a break, and for test runs it is similarly possible to transfer the essential load to the standby supply without a break, while leaving the nonessential load connected to the normal supply. It should be noted that this advantage can only be gained if the normal supply circuit breaker connects to the section of busbar that supplies the nonessential load. (See Figs. 2.2 and 2.3.) These procedures require paralleling for short periods only and the agreement with the network operator usually limits the duration to 5 min.

The technical requirements are set out in Engineering Recommendation G59/1 and its supporting document Engineering Technical Report 113, both published by the Electricity Association.

Features Required on the Generating Set

The generating set should have been constructed with parallel running in mind. The coupling between the engine and alternator should be sufficiently robust to withstand an attempt to synchronize in phase opposition without suffering damage; a shear pin is sometimes provided between the alternator and the coupling. The alternator poles should be provided with pole face damper windings to prevent phase swinging.

As the set will be feeding into an infinite system the engine governor will have no control over the speed and the voltage regulator will have no control over the voltage. Instead the governor is used to control power (kW_e) and voltage regulator to control power factor ($cos\phi$). For synchronizing purposes the engine governor and the voltage regulator will require remote adjusting facilities.

The Supply Authority's Requirements

For reasons connected with the safety of their operating and maintenance personnel the network operator will require electrical protection to ensure that their distribution network is not energized from the standby supply. For long-term continuous paralleling the protection required is complex but for the short-term occasional paralleling applicable to standby installations the protection required will usually consist of under- and overfrequency, under- and overvoltage, overcurrent, earth leakage, and a 5-min timer, in addition the network operator may require protection against neutral voltage displacement. The settings of the relays required for paralleling will be determined by the network operator.

The purpose of the under- and overfrequency and voltage protection is to detect a loss of normal mains and to prevent the standby generator back-feeding a part of the supply authority's distribution system (islanding). The purpose of the neutral voltage displacement protection is to prevent the standby generator back-feeding an earth fault on the supply authority's system. For some installations neutral voltage displacement protection may be expensive or impracticable, and in such cases it is likely that agreement could be reached on the use of some other parameter such as rate of change of frequency, rapid phase angle change, or unbalanced voltages.

Full tests on the protective equipment must be undertaken and recorded by the installer to the satisfaction of the network operator. Where the standby supply is connected to a high-voltage system the network operator has a duty to witness the tests; where it is connected to a low-voltage supply the operator may, at its discretion, wish to witness the tests.

The loaded generator is unlikely to produce a pure sine wave of voltage and when it is connected to the normal mains it will cause harmonic currents to flow. As with any distorting load connected to the U.K. system, the harmonic currents must be within the limits set by Engineering Recommendation G.5/4 (ER G.5/4) published by the Electricity Association. More is written about ER G.5/4 and harmonics in Chap. 4 of this book.

The Low-Voltage Neutral Connections

Where a low-voltage standby supply is intended to run in parallel with the normal supply, the two neutrals may be solidly connected. For TN-C-S (PME) systems this is the simplest arrangement as the standby supply neutral may be solidly earthed and connected to the normal supply neutral; the changeover from one supply to the other is then effected by means of triple-pole contactors or circuit breakers. However this will probably not be practicable, it is likely that the resulting triplen harmonic current flow in the neutral will exceed the limits set by ER G.5/4 and some other procedure must be adopted.

One procedure is to provide a single-pole neutral contactor in the connection to the generator star point, and to arrange for this to be open when running in parallel. With this arrangement the standby supply star point is not used and the system relies upon the normal supply for single-phase loads, for earthing, and for triplen harmonic currents; all the zero-sequence currents flow from the normal supply phase conductors and return additively in the neutral conductor. Harmonic current flow in the phase conductors of the normal supply must not exceed the limits set by ER G.5/4. The generator manufacturer must provide sufficient information for the network operator to be able to calculate the harmonic current flow before permission for paralleling is given.

There must be electrical interlocks to ensure that:

- It is not possible to connect the two supplies in parallel unless the neutral contactor is open.

■ It is not possible to use the standby supply independently unless the neutral contactor is closed.

An alternative procedure is to install a reactor connected between the generator star point and the supply neutral. The reactor is to have sufficient reactance to limit the triplen harmonic current flow to within the limits set by ER G.5/4. This arrangement is discussed in ER G.59/1 and in ETR 113 but it is more complex and expensive than the preceding procedure and would seem to have limited application for single sets.

Earthing Low-Voltage Supplies

When the normal and standby supplies are running in parallel the neutral connection will usually be taken from the normal supply and, depending on the earthing arrangements and the agreement with the network operator, there may be a requirement for an earthing contactor or for an additional pole on the neutral contactor mentioned in an earlier paragraph. For multiple earthed systems, including PME, the neutral and earth connections are combined and a single pole neutral contactor is used. Figure 2.17 illustrates the electrical connections. For single-point earthed systems such as TN-S or TT an additional pole is required on the neutral contactor to avoid a second earth connection to the normal supply neutral. Figure 2.18 illustrates the electrical connections. If a separate earthing contactor is used, electrical interlocks, as required for the neutral contactor, must be provided.

As was the case for the diagrams related to alternative connections, Figs. 2.17, 2.18, and 2.19 have been derived from diagrams appearing in ETR 113. Figure 2.17 provides a parallel path for the neutral current, the undesirable consequences of such an additional path have already been discussed at the end of the section titled "Neutral Connections for Single Sets Not Intended to Run in Parallel with the Normal Supply" in connection with Fig. 2.7.

Earthing High-Voltage Supplies

For high-voltage systems the network operator may use solid, resistor, reactor, or arc suppression coil earthing methods and a second earthing point will not normally be allowed. When a high-voltage standby supply is operating independently its star point must be connected to earth, usually through an earthing resistor, but when it is operating in parallel with the normal supply, the earth connection must be opened, for which purpose an earthing contactor is used. Figure 2.19 illustrates the electrical connections. With this arrangement the standby supply star point is not used and the system relies upon the normal supply for earthing.

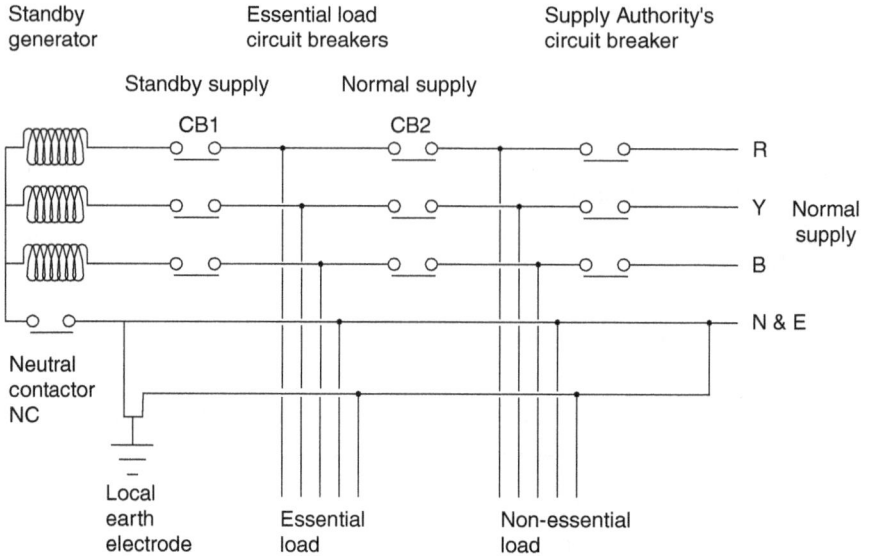

Synchronizing facilities are provided across each of the essential load circuit breakers.

Figure 2.17 Typical connections and switching status table for low-voltage parallel running when the normal supply is multiple earthed (PME, TN-C-S, or TN-C).

Overcurrent Protection of the Standby Supply

Behavior of Generators under Fault Conditions

When a generator experiences a fault, a large current, dependent upon the subtransient reactance (X''), flows for a short time, this will rapidly decay to a value dependent upon the transient reactance (X') and at the same time the voltage regulator will start to increase the excitation. After a short time the current will have stabilized to a

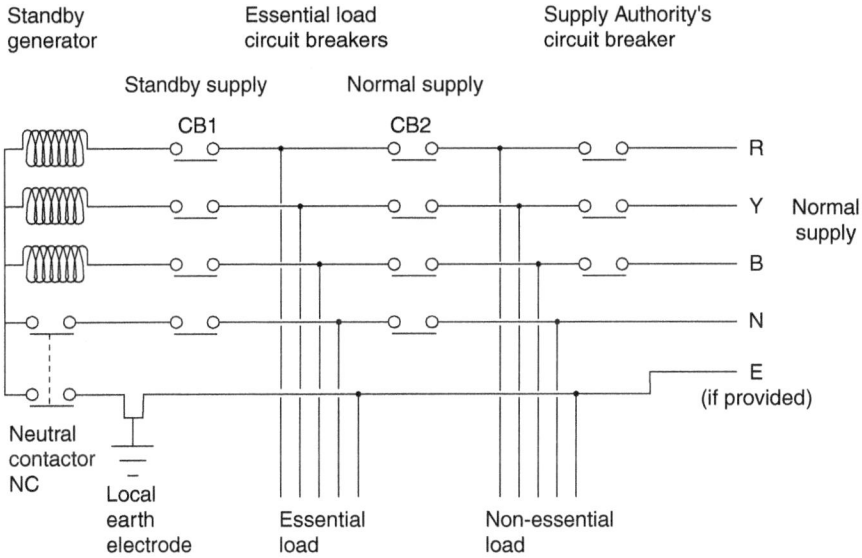

Device	CB1	CB2	NC
Essential load from normal supply	O	I	O
Essential load from standby supply	I	O	I
The two supplies in parallel	I	I	O

Synchronizing facilities are provided across each of the essential load circuit breakers.

Figure 2.18 Typical connections and switching status table for low-voltage parallel running when the normal supply is single point earthed (TN-S or TT).

steady-state value determined by the synchronous reactance; this is the steady-state short circuit current which is available for relay operation (the prospective fault current). The time constants of these decrements depend on the generator size and design, the subtransient time constant is measured in milliseconds, the transient time constant in tens of milliseconds and the synchronous time constant in tenths of a second.

For satisfactory fault clearance it is important that each generator is provided with an effective excitation system. The simplest excitations systems use a shaft-mounted rectifier, an ac exciter, and

Standby   Essential load   Supply Authority's
generator   circuit breakers   circuit breaker

Standby supply   Normal supply

CB1   CB2

R

Y   Normal
supply

B

E
Earthing   (if provided)
contactor
EC

Earthing
resistor

Essential   Non-essential
load   load

Local
earth
electrode

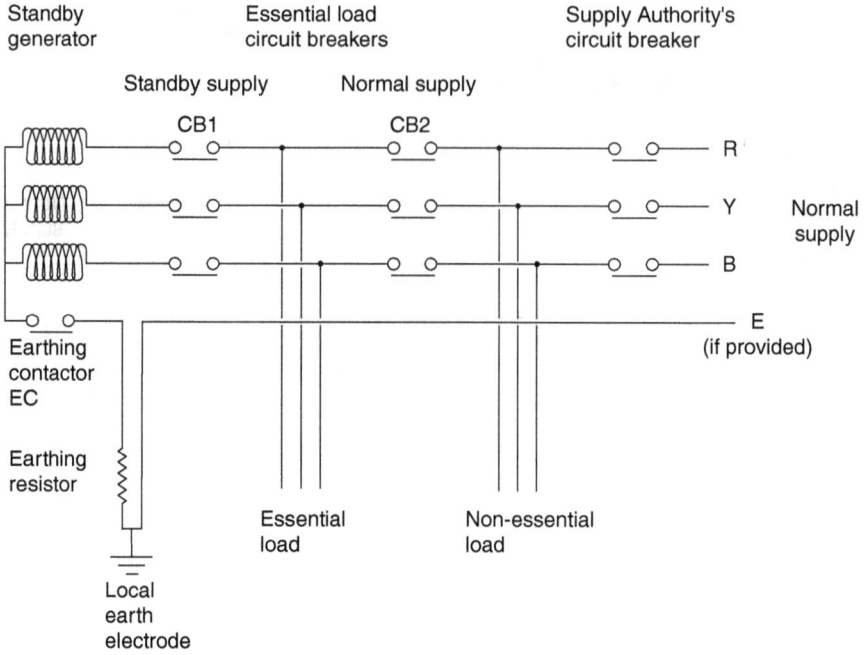

Device	CB1	CB2	EC
Essential load from normal supply	O	I	O
Essential load from standby supply	I	O	I
The two supplies in parallel	I	I	O

Synchronizing facilities are provided across each of the essential load circuit breakers.

Figure 2.19 Typical connections and switching status table for high voltage parallel running.

a pilot exciter with permanent magnet excitation (see Fig. 1.4). Alternatively, instead of a pilot exciter, power may be derived from the generator output terminals (see Figure 1.5); with this arrangement, under short circuit conditions, there will be a loss of exciter field unless a system to maintain it is included. One system uses current transformers to maintain excitation during a fault. Another system derives power from a dedicated third harmonic winding in the stator.

The steady-state short circuit current available from an ac generator is normally of the order of three times the rated current.

Protection of the Generator

Small sets up to say 75 kW may be protected against overload and fault conditions by fuses or by molded-case circuit breakers, and this may well be the only form of protection used.

For sets above 75 kW the basic form of protection against overloads is to provide, for each generator, a three-phase overcurrent relay with an inverse time characteristic and having sufficient delay to allow downstream protection to operate. To allow positive discrimination between two relays it is usual to ensure that the two characteristics provide a time difference between 0.3 and 0.4 s. The shorter time would be appropriate for installations using modern electronic relays (for greater accuracy) and quick operating circuit breakers such as those using vacuum tubes.

For sets above 75 kW the form of protection against faults in the distribution system is to provide three instantaneous high set overcurrent elements arranged to operate a timer adjustable up to 5 s. The time delay is to allow any downstream protection to operate before taking the somewhat drastic step of shutting down the entire standby supply. Typical connections are indicated by Fig. 2.20.

It is worth noting that if the current transformers measuring overcurrents are mounted at the neutral ends of the stator winding, they will react to any internal phase-to-phase or phase-to-earth faults. However, except for large sets this is not usually possible and internal faults are monitored by other systems.

The following additional protection systems may be considered:

- *Restricted Earth Fault Protection.* In this system the currents entering and leaving the stator are summed and the sum is applied to a relay. If the sum is not zero an internal winding fault is indicated and the generator should be deexcited and the engine shut down; since no discrimination with downstream protection is involved, no time

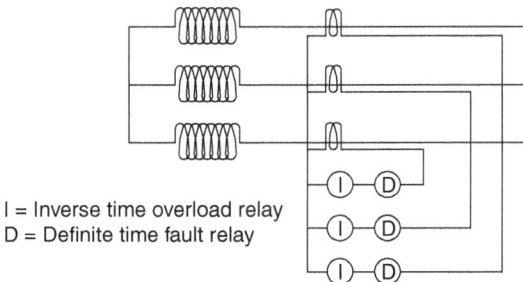

I = Inverse time overload relay
D = Definite time fault relay

Figure 2.20 Typical connections to provide inverse time-delayed overload and definite time fault protection.

delay is necessary. Downstream faults do not have any effect on the relay, hence the use of the word *restricted* in the name. The system requires four current transformers, one in the neutral which can be mounted in the terminal box or in the switchgear, and three current transformers in the phases which are usually mounted in the switchgear. The area protected lies between the three-phase current transformers and the neutral transformer. The system is applicable to sets of above say 100 kW. It is sometimes known as a *balanced current system*. Typical connections are indicated by Fig. 2.21.

- *Differential Current Protection.* In this system the currents at the start and finish of each phase of the stator windings are compared. If any of the phases is not balanced, this indicates an internal fault and the generator should be deexcited and the engine shut down; since no discrimination with downstream protection is involved, no time delay is necessary. The system requires the three ends of the stator winding to be brought out of the stator and into the neutral terminal box and the installation of three current transformers in the box, the other three current transformers and the relays are usually mounted within the associated switchgear. This system is applicable to large sets, say above 1.5 MW. The area protected lies between the two sets of current transformers and for this reason the term *unit protection* is sometimes used. The term circulating current system is also sometimes used. Typical connections are indicated by Fig. 2.22.

- *Standby Earth Fault Protection.* This system is applicable to high-voltage generators which are resistance earthed. In this system the current flowing to earth from the star point is monitored and is applied to a relay. Any current flow indicates a fault and should operate an inverse time-delayed trip. The system is unrestricted and the time delay is required to allow any downstream protection to operate.

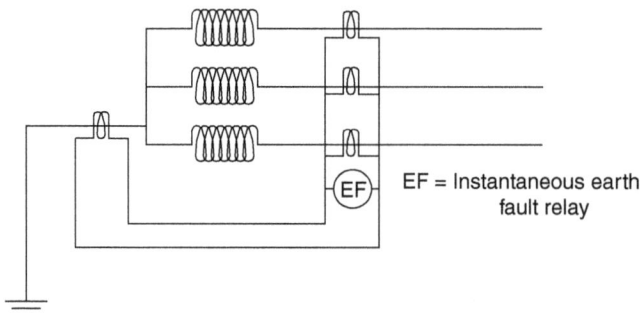

EF = Instantaneous earth fault relay

Figure 2.21 Typical connections to provide restricted earth fault protection.

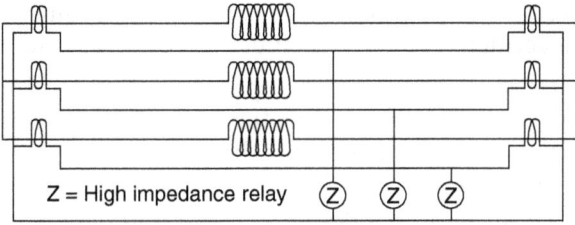

Figure 2.22 Typical connections to provide differential current protection.

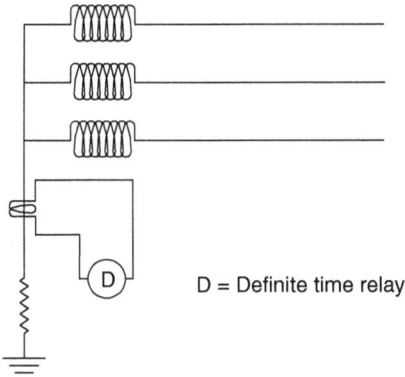

Figure 2.23 Typical connections to provide standby earth fault protection.

The generator should be deexcited and the engine shut down if the fault persists after the expiry of the time delay. This protection is a useful supplement to the restricted earth fault or differential current protection, it also protects the earthing resistor against continuous loading which would have disastrous consequences. Typical connections are indicated by Fig. 2.23.

- *Reverse Power Protection.* This system is needed for any generator which is required to run in parallel with another supply or another generator. A reverse power situation indicates that the generator is acting as a motor and is driving the engine, an undesirable situation. Under reverse power conditions the generator circuit breaker should be opened, it is not important to shut down the engine immediately but it must not be left running unloaded for longer than say 15 min, it depends on whether the generator is attended or not. It is usual for the relay to include a short time delay to prevent any transient effects, particularly during synchronizing operations, from tripping the circuit breaker.

- *Unbalanced Load Protection.* This system is applicable where there is concern that the generator may be running for long periods on a load which is unbalanced between phases. It is not normally required and is mentioned here because ISO 8528 includes a reference to it.

■ *Voltage Restrained Overcurrent Relays.*   The opening paragraphs of this section explain the behavior of a generator under fault conditions. When running from a standby generator the low value of short circuit current can lead to problems with fault clearance. The voltage restrained relay has two time/current characteristics and may help to resolve the problems. Under conditions of normal voltage the relay has a long time characteristic which is intended to ensure that the generator remains on line for as long as possible. If conditions of low voltage arise during fault clearance, the relay operates to a standard inverse characteristic and, hence, trips in a predictably shorter time.

A Note on Current Transformers and Relays

Current transformers used for protection purposes must be of the correct class for the duty. It is important that their cores do not saturate at the maximum currents expected, it is also important, for differential and earth fault systems, that the ratio and secondary phase angle are sufficiently accurate to ensure proper operation.

Until the 1980s most protection relays were electromechanical and based on the induction disc or cup, the principle having remained in use for many years. Current adjustment was by a "plug setting" and a tapped autotransformer, and time adjustment was by a "time multiplier" which determined the angle through which the disc had to rotate to close its contacts. The parts required precision manufacture and skilled assembly and, over a long period of time, as electronic techniques improved, the electromechanical relays became obsolescent and were superseded by electronic versions. For this reason all relays are now electronic in operation and have a very different appearance. Similar comments have already been made in this book in connection with voltage regulators and speed governors.

The modern relays perform the same functions as their predecessors and the same terminology is used, adjustments are made in a different way to achieve the same result. Electronic relays are more accurate than the electromechanical devices, overshoot is less because there are no moving parts, and the current transformer burdens are less. The relays are more versatile and provide, within a single unit, a choice of characteristics such as standard inverse, very inverse, extremely inverse, or definite time. The final output circuits may use electromagnetic relays with metallic contacts, which are better suited to the inductive loads likely to be encountered than are electronic components.

In the previous paragraph the characteristics described as standard inverse, very inverse, and extremely inverse are the operating characteristics described in BS EN 60255. The characteristics are inverse with a definite minimum time and are known in full as inverse definite minimum time characteristics, sometimes abbreviated to IDMT.

Protection of the Distribution System

In a normal situation the overload relay characteristic of the supply would appear on the time grading graph to the right of and above the characteristics of the distribution system. Any fault on the distribution system would be cleared, enabling the healthy parts of the system to continue in operation. However, the prospective fault current obtainable from the standby supply will be much less than that obtainable from the normal supply, and will probably be of the order of three times the rated current. On the protection time grading graph the effect is to move to the left, sometimes drastically, the supply characteristic. This characteristic may intersect some of the existing operating characteristics, thus lengthening their operating times or, worse, may leave them isolated in a high current section of the graph where the standby supply cannot operate.

If the standby supply feeds a number of loads through dedicated changeover contactors or circuit breakers there should be no problem in providing protection because the distribution system is in effect duplicated. Where a small standby supply feeds a large distribution system and uses the same switchboard as the normal supply, it will be unlikely to be able to clear faults in the main distribution system and a cable fault near the power source may not be cleared, thus rendering the standby supply inoperative, until it can be restored by manual switching or repair. Such an event is unlikely and the risk may be acceptable; provided that the generator rating is large in relation to the final circuit protective devices, disconnection times complying with BS 7671 should be achievable for the final circuits. The alternative is to connect the standby supply to a downstream point where the protection will be set at a lower level.

Switchgear

For the smaller and simpler generating sets, the switchgear may comprise a single molded-case circuit breaker mounted on the generating set base frame. For larger sets the switchgear is installed separately, and for multiple-set installations the switchgear will comprise several cubicles with busbars and interconnecting wiring. Sometimes, such switchgear is installed in the engine room, but this is an extremely noisy location in which to operate switchgear under the stressful conditions applying after a power or equipment failure. There are advantages to installing the switchgear in a separate room away from the noise and high ambient temperature of the engine room; if there are problems to be resolved it will certainly be easier to discuss the matter or to exchange technical information.

The Cable Connecting the Switchgear to the Generator

The cable entering the generator terminal box should be a flexible cable and should be so arranged and run that it does not come under stress due to vibration or movement of the set. If the cable is not flexible or is not properly installed, it may restrict the freedom of movement of the set, affect the performance of the antivibration mounts, and cause vibration problems.

If the length of the cable run between the switchgear and the generator is short, say not exceeding 10 m, the entire run can be in flexible cable. If the run is longer a cable change box may be installed so that the majority of the cable is of standard distribution type and only a short length of flexible cable is necessary.

For some small sets the main circuit breaker is mounted on a framework fixed directly to the generating set base frame, the manufacturer providing the interconnections between the circuit breaker and the generator. For such an arrangement the interconnections should be in flexible conductors and the cable taking power from the circuit breaker may be of standard distribution type.

For fault protection purposes, this cable may be regarded as an extension of the stator windings, and if the switchgear is remote from the generator some additional protection may be desirable. The probability of a fault developing in this cable is small but the condition can be detected by installing a restricted earth fault protection system as described in the previous section of this chapter. In addition to providing cable fault protection, this system provides protection against internal winding faults and limits the resulting damage to the machine.

For installations of more than one set, the restricted earth fault protection should be provided on each set.

Low-Voltage Switchgear

For low-voltage installations, the switchgear should be a type tested assembly (TTA) or a partially type tested assembly (PTTA) complying with BS EN 60439-1. The switchgear and controlgear components within the switchboard should comply with BS EN 60947. The enclosures should protect the equipment against the ingress of solid foreign objects, and persons against touching hazardous parts, to one of the degrees of protection in BS EN 60529 which is invoked by BS EN 60439-1.

High-Voltage Switchgear

For high-voltage installations, the switchgear should comply with BS EN 60298. The enclosures should protect the equipment against the ingress of solid foreign objects, and persons against touching hazardous parts, to one of the degrees of protection in BS EN 60529. (BS EN 60298 invokes BS EN 60694 which in turn invokes BS EN 60529).

Prospective Fault Current Level

For a set providing an alternative supply and not arranged to run in parallel with the normal supply, the prospective fault level will be the higher of the two supplies, which will almost certainly be that of the normal supply. For sets running in parallel with the normal supply, the prospective fault current will be the sum of the two fault currents. The mechanical and thermal stresses imposed on the busbars are proportional to the square of the current, it follows that a modest additional current from a generator can have a significant effect within the switchboards.

Information to Be Displayed at the Switchboard

Single-Set Installations The generator panel of a single set, not intended to run in parallel with another set or the normal supply, is required by ISO8528 to display the following information:

Voltage, with provision for indication of phase-to-phase and phase-to-neutral voltages

Current, with provision for indication of each phase

Other indications which should be considered includes:

The frequency of the ac output

Operating hours, that is, a cumulative hours-run counter

For large sets only, indication of kW, kVAr, power factor, and cumulative kWh.

Multiple-Set Installations Each generator panel of a multiple-set installation not intended to run in parallel with the normal supply should display the following:

Voltage, with provision for indication of phase-to-phase and phase-to-neutral voltages

Current, with provision for indication of each phase

kW indication for each set

For manual synchronizing, the switchboard should include a synchro-scope, capable of being connected to any generator and indicating any difference frequency and the phase relationship between the busbars and the generator before paralleling. It is usual to include synchronizing lamps or a zero voltmeter and a check synchronizer as back-up features.

For automatic synchronizing, the switchboard should include an automatic synchronizer feeding signals to motor operated adjustments of the voltage and speed regulators of individual sets.

Other indications which should be considered include:

The frequency of any set or of the busbars to be available for display when required

Operating hours, that is, a cumulative hours-run counter for each set

Indication of kVAr, power factor, and cumulative kWh for each set

Indication of total kW, total kVAr, and the standby supply power factor

Sets Intended to Run in Parallel with the Normal Supply   The information relating to multiple sets applies, with the addition of the following:

Voltage, provision for indication of the busbar voltage which, when the standby supply is shut down is the normal supply voltage.

The synchronizing features should operate between the busbars and the normal supply, and between the busbars and the standby supply, thus enabling synchronizing in either direction. For multiple-set installations the synchronizing features should also operate between any individual set and the busbars.

Other indications which should be considered include:

The frequency of any set or of the busbars to be available for display when required

Operating hours, that is, a cumulative hours-run counter for each set

Indication of kVAr, power factor, and cumulative kWh for each set

Indication of total kW, total kVAr, and standby supply power factor if more than one generating set is installed

The Synchronizing Facility

A switchboard which provides a paralleling facility will include one set of synchronizing equipment which the user is able to connect between any incoming supply and the busbars. Selection is by a switch, plug and sockets, or any other means, it must not be possible to select more than one incoming supply.

There must be provision for occasionally connecting to "dead busbars" when, obviously, synchronizing is impossible and the control system must take care of this condition. It arises when a single set or the first of multiple sets is started and connected to "dead busbars."

For manual synchronizing the synchroscope is arranged to be easily visible from each generator panel, and each panel includes raise/lower controls for speed and excitation so that the operator can make fine adjustments while watching the synchroscope. It is likely that the electromechanical synchroscope will be replaced by fully electronic versions, but clear visibility is essential. It is usual to include a check synchronizer in the control system; the check synchronizer prevents circuit breaker closure unless the voltage, frequency, and phase relationship are within limits.

For automatic synchronizing the control system takes care of the adjustments to voltage, frequency, and phase relationship, and closes the circuit breaker when conditions are within limits.

Test Load Facility

Consideration should be given to including in the switchboard a switched outlet to which a test load can be connected when required. The circuit should be fused or otherwise protected and fully rated for the output of one set. For a multiple-set installation this will usually be adequate, but for a large important installations the rating could be increased. This facility is useful after major work has been undertaken on a set and avoids the need for temporary connections within the switchboard in order to connect a test load. The connection of a permanently installed test load to this circuit is discussed in Chap. 3.

Standard Reference Conditions

The standard conditions for switchgear are specified in ISO 8528 as:

Altitude.   Not exceeding 2000 m above sea level. *Note*: For high-voltage switchgear BS EN 60694 sets the maximum at 1000 m above sea level.

Air temperature.   Maximum not exceeding 40°C and 24-h average not exceeding 35°C.

Relative humidity.   Not exceeding 50 percent at 40°C. *Note*: For high-voltage switchgear more complex rules are given in BS EN 60694.

It should be noted that, at altitudes above the reference level, the reduced density of the air leads to a reduction of the dielectric strength and of the cooling effect, and special considerations are demanded.

Bibliography

British and European Standards

BS 7430—Code of practice for earthing
BS 7671—Requirements for electrical installations (the wiring regulations) (IEC 60364 covers the same subject)
BS EN IEC 60255—Electrical relays
BS EN IEC 60298—A.C. metal enclosed switchgear and controlgear for rated voltages above 1 kV and up to and including 52kV
BS EN IEC 60439—Specification for low-voltage switchgear and controlgear assemblies
Part 1—Type tested and partially type tested assemblies
BS EN IEC 60529—Specification for degrees of protection provided by enclosures (IP code)
BS EN IEC 60694—Common specification for high-voltage switchgear and controlgear standards
BS EN IEC 60947—Low voltage switchgear and controlgear
Part 1—General rules

Other Documents

The Electricity Supply Regulations 1988. Statutory Instrument 1988 No. 1057. Her Majesty's Stationery Office of London, United Kingdom.
Engineering recommendation G59/1—Recommendations for the connection of embedded generating plant to the public electricity suppliers' distribution systems. Electricity Association of London, United Kingdom.
Engineering technical report No.113—Notes of guidance for the protection of embedded generating plant up to 5 MW for operation in parallel with public electricity suppliers' distribution systems. Electricity Association of London, United Kingdom.

Additional Information Relating to the Standby Supply Installation

Introduction

This chapter is concerned with various matters which do not fit comfortably into either of the two previous chapters. The calculation of engine and generating ratings is a somewhat tortuous exercise and has been left until this late stage in the book so that the reader has the benefit of text appearing in earlier parts of the book. The components of the load current, characteristics of particular loads, vibration, and noise are discussed.

In this chapter the expressions kWm and kWe are used to denote mechanical and electrical power respectively.

Sizing the Engine and Generator

The Power Rating of the Engine

The factors to be considered in sizing the engine are:

- The class of power rating specified for the generating set (continuous power, prime power, or limited time running power)
- The frequency performance class G1, G2, G3, G4, or as otherwise specified (see Table 1.1)
- The maximum load (kWm) to be supplied
- The maximum step of load (kWm) to be applied

The engine must be capable of supplying the total required mechanical load to the generator. This will be higher than the electrical load which the set is intended to supply, the additional loads to be considered being:

- The losses within the generator, which may be 5 or 6 percent
- The radiator fan if it is not engine driven
- Any electrically driven ventilation fans, engine room lighting, etc., that are not included in the load to be supplied
- Any engine auxiliaries such as pumps that are driven by electric motors
- Switchgear and distribution losses, which should be small (probably less than 0.5 percent) but should not be forgotten

It will be seen that the total of these loads and losses could be significant and must be considered in the engine power rating.

The Step Loading Ability of the Engine

Only naturally aspirated diesel engines are capable of accepting full mechanical load in one step, most modern engines are turbocharged. The maximum step load acceptable is very much dependent upon the engine brake mean effective pressure, and manufacturer's advice must be sought. ISO 8528 includes guide values which appear elsewhere in this book. Typically, modern engines will accept a 60-percent step of load and a portion of the total load has to be separated and the power supply to it delayed in order to keep within the limit.

When load is applied to the engine there is an immediate deceleration and, until the governor is able to respond, some energy is extracted from the inertia of the engine, flywheel, and generator rotating system. The governor responds by moving the fuel rack to increase power and normal speed is restored.

The kVA Rating of the Generator

The factors to be considered in sizing the generator are:

- The class of power rating specified for the generating set (continuous power, prime power, or limited time running power)
- The voltage performance class G1, G2, G3, G4, or as otherwise specified (see Table 1.1)
- The maximum load (kWe) to be supplied
- The maximum reactive load (kVAr) to be supplied
- The maximum step of reactive load (kVAr) to be applied

The generator must be capable of supplying the total required electrical load to the distribution system. This will be equal to the power rating of the engine as derived above less the losses within the generator (5 or 6 percent).

While the basic engine rating is expressed in kWm the basic generator rating is in kVA, which is the phasor sum of kWe and kVAr. The relationship is defined by the well-known equations:

$$\text{kW} = \text{kVA} \times \cos \phi \qquad (3.1)$$

and
$$\text{kVAr} = \text{kVA} \times \sin \phi \qquad (3.2)$$

The Effect of UPS Loads on the Generator kVA Rating

Most standby power generators provide power to one or more uninterruptible power supply systems which may use six-pulse rectifiers as their input modules. As described in Chap. 4, a six-pulse rectifier generates harmonic currents of the orders 5, 7, 11, 13, etc., the theoretical total harmonic distortion being 30 percent. In practice it is usually higher. In passing through the generator these harmonic currents will cause voltage distortion which is undesirable for many reasons. If the total harmonic distortion of the generator voltage is to be limited to say, 10 percent, the total six-pulse rectifier load on a generator should not exceed 30 or 40 percent of its kVA rating, depending on whether it has high or low subtransient reactance. A precise calculation requires the use of data which will be available to the generator manufacturer.

The Step Loading Ability of the Generator

Generators experience no difficulty in accepting step loads up to 100 percent or more of their rating. The resulting transient voltage drop, however, has to be considered as it must be within the limits defined by the voltage performance class (G1, G2, G3, or G4).

When load is applied to a generator, there is an immediate drop in output voltage due to the stator subtransient reactance (X'') and resistance (R). The voltage regulator will respond quickly, but the exciter and rotor fields are highly inductive and, owing to their time constants, there is delay in increasing the excitation of the machine. The generator rotor will also suffer a loss of speed which will further reduce the voltage. For self-excited machines (in which the exciter field is energized from the generator output) the voltage drop at the generator terminals is transferred to the exciter field, still further increasing the drop. It is likely that the output voltage drop will be increased by armature reaction before excitation begins to increase. It can be seen that the generator voltage control loop is complex and the calculation of voltage dip on application of load is best undertaken by the machine designer. The drop cannot be less than the product of the impact kVAr and the subtransient reactance.

The subtransient reactance (X'') is due to the stator leakage flux, in other words, that part of the stator flux which does not link with the rotor windings. The leakage flux originates in the stator winding overhangs and in the stator slots themselves. The stator windings will also have some resistance which is inseparable from the reactance but the value is low compared with the reactance and it is has little effect.

The Voltage Drop Caused by Step Load Application

For normal loads having a power factor of say 0.8, the voltage drop on application of generator rated kVA is likely to be 14 to 16 percent for a self-excited machine (in which the exciter field is energized from the generator output), and 11 to 13 percent for a machine having a permanent magnet pilot exciter. The limitation of engine step loading usually precludes the application of full rated kVA to the generator but the voltage drop may be assumed to be proportional to the applied kVA.

The Voltage Drop Caused by Motor Starting

Starting of squirrel cage induction motors is a common cause of voltage dips and the most common method of starting is to connect direct on line. The direct on-line starting current (impact current) is likely to be seven times the rated current for motors up to 200kWm, and six times rated current for larger motors, but there is wide variation. The power factor will be low, of the order of 0.15. The voltage drop on application of generator rated kVA is likely to be 18 to 20 percent for a self-excited machine, and 14 to 16 percent for a machine having a permanent magnet pilot exciter. The impact kWe will typically be of the order of 120 percent of the running kWe, but special motors such as those designed for high torque starting will take a higher impact kWe.

As the motor accelerates the current taken reduces and its power factor increases. The locus of the current phasor is a semicircle as shown in Fig. 3.1.

During acceleration the kWe demand rises to a maximum and then reduces to the running power. Provided that the motor is not coupled to a high inertia load the peak power will be transient and of short duration. For low-cost installations use is sometimes made of the power which the engine has in reserve for governing purposes, but this power is not intended for this purpose and the practice is not recommended. The peak

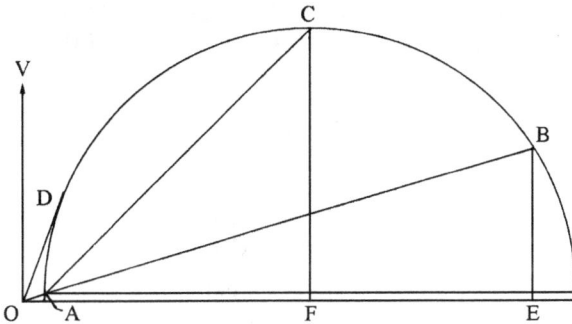

V - voltage phasor   OC - current at maximum accelerating kWe
OA - stator magnetizing current   OD - rated running current
OB - starting (impact) current   EB - starting (impact) kWe
FC - maximum accelerating kWe input

Figure 3.1 Typical circle diagram of a squirrel cage induction motor.

power (kWe) required will depend on the design of the motor but it will be several times the rated power (kWm) output.

For star/delta starting the considerations are similar to those applying to direct on line starting but the applied voltage and the current are effectively reduced by a factor of $1/\sqrt{3}$ and the power by a factor of 1/3. There will be less leakage flux and the impact power factor will be higher at say, 0.4. There will be two impact loads, the first being determined by the motor characteristics, the second by the time setting for the star/delta changeover.

For auto/transformer starting similar reasoning applies.

Example of Engine and Generator Sizing

Manufacturers have a wealth of experience and information at their disposal and are well placed to calculate the performance of the engine/ generator assembly. The planner or user usually has neither the experience nor the detailed information. The following method of estimating performance is suggested as a guide, it should not be regarded as a definitive method.

Assume that a generating set is to operate under normal reference conditions and is required to supply a mixed load of lighting, air conditioning, small pumps, fans, etc., of 240 kW at a power factor of 0.8. The generating set is to have a prime power rating. The supply voltage is 400 V three phase. In addition there are two direct on-line started motors:

Motor A is rated at 60 kW with a full load current of 105 A at 0.89 power factor. At starting it takes 735 A at a power factor of 0.15, and the peak accelerating power is 150 kWe at a power factor of 0.6.

Motor B is rated at 70 kW with a full load current of 123 A at 0.89 power factor. At starting it takes 861 A at a power factor of 0.15, and the peak accelerating power is 175 kWe at a power factor of 0.6.

In this example the following assumptions are made:

The generator efficiency is 94 percent.

The generator has a permanent magnet pilot exciter and the volt drop on application of full-rated low-power factor kVA is 11 percent.

The maximum permissible voltage drop is 12 percent.

The load as stated includes all the electrically driven engine auxiliaries and the charging of any UPS batteries.

The loading sequence may be set out in tabular form, the completed table appearing as Table 3.1. It includes a great deal of information and an explanation of how it is compiled follows. The left-hand section entitled "Applied load" is compiled first, and it will be found that, for each line of this part of the table, two items of information are available. For the mixed load, the kWe and cos ϕ are known; for the starting and running conditions of the motors, the current and cos ϕ are known; and for the peak accelerating power, the kWe and cos ϕ are known. From the two known figures in each line it is possible to calculate the other three values.

The last line of the table indicates the load at the end of the loading sequence, and two values can again be extracted from the preceding figures. The figure for kWe is the sum of the mixed load and the two motor running loads, that is, 240 + 76 + 65 = 381, and for kVAr is similarly 180 + 39 + 33 = 252. From these two figures the other three values can be calculated; neither current nor kVA should be added arithmetically because they include phasor quantities.

The right-hand section entitled "Cumulative Load" can now be compiled. The mixed load is the first to be applied and the cumulative load figures are the same as for the applied load. The second line down is the motor B starting transient condition and the kWe figure is the sum of the mixed load and the motor starting load, that is, 240 + 89.5 = 329.5, and the kVAr figure is similarly 180 + 590 = 770. The kVA figures cannot be obtained by arithmetic addition and have to be calculated if they are required.

The third line down is the motor B accelerating transient condition and the kWe figure is the sum of the mixed load and the motor accelerating load, that is, 240 + 175 = 415, and the kVAr figure is similarly 180 + 233 = 413.

The fourth line down is the motor B steady state running condition and the kWe figure is the sum of the mixed load and the motor running load, that is, 240 + 76 = 316, and the kVAr figure is similarly 180 + 39 = 219.

TABLE 3.1   Standby Supply Loading Sequence

Description	Applied load					Cumulative load		
	Amps.	cos φ	kWe	kVAr	kVA	kWe	kVAr	kVA
Mixed load	433	0.8	240	180	300	240	180	300
			Motor B					
Starting (transient)	861	0.15	89.5	590	597	329.5	770	837
Accelerating (transient)	421	0.60	175	233	291	415	413	585
Running	123	0.89	76	39	85	316	219	385
			Motor A					
Starting (transient)	735	0.15	76.4	504	510	392.4	723	823
Accelerating (transient)	361	0.60	150	200	250	466	419	627
Running	105	0.89	65	33	73	381	252	457
Final running load	660	0.83	381	252	457	381	252	457

The fifth, sixth, and seventh lines apply to motor A and may be similarly completed. The figures in the seventh line should agree with the final running load obtained in the left-hand section of the table.

The table tells us, from the applied load kWe column, that the largest step load is 240 kWe and from the cumulative load kWe column, that the largest cumulative load is 466 kWe, both as seen by the generator. It confirms that the final or continuous load is 381 kWe, again as seen by the generator.

The table can be expanded to include any relevant features of the loading sequence. If, for instance, the installation includes uninterruptible power supplies, their batteries will require recharging after a power failure; this may increase their power demand by 25 percent for a period of time which will depend on how long they were deprived of power. Sometimes the battery charging is delayed until the loading sequence is complete and the generating set is in a steady-state mode of operation.

It will be noted that motor B (70 kW) has been started before motor A (60 kW). If it is permissible larger motors should be started first because the engine is then on a low load. If motor A had been started first the maximum cumulative load seen by the engine would have been greater than 466 kWe.

The Engine Rating

The continuous load is 381kWe which, taking into account the alternator efficiency, will require 405 kWm from the engine. The maximum cumulative load seen by the generator is 466 kWe when motor A is accelerating, which will require 496 kWm from the engine. It follows that the maximum or prime power required from the engine will be 496 kWm, and the continuous power will be 405 kWm, or 81.6 percent of the maximum. The average power over a 24-h period will depend on the duration of the initial sequence and will be a little over 81.6 percent.

The maximum step load applied to the generator is 240 kWe which will require 255 kWm from the engine. This is 51.5 percent of the maximum rating, an acceptable figure since most engines will accept a 60-percent step load.

For some applications a simpler table will suffice. Table 3.1 includes some redundant information, which may, however, be useful as it assists in checking the arithmetic used in compiling the table. The full information also enables the reader to see all the changes that occur during the loading sequence.

Unless a safety margin has been allowed in the loads given, it would be wise to increase these ratings by say 10 percent. ISO 8528 does not allow for any overload rating, whereas ISO 3046 requires engines to be capable of supplying 10 percent overload for 1 h.

The Generator Rating

The continuous load is 457 kVA.

The maximum step load is 597 kVA when motor B is started.

If a generator rated at 500 kVA is used the volt drop on application of motor B would be:

$$\text{Drop} = (597 \div 500) \times 11\% = 13.2\%$$

To restrict the volt drop to 12 percent would require a rating of:

$$500 \times 13.2 \div 12 = 550 \text{ kVA}$$

The preceding calculations are an estimating procedure which will give approximate results. The generating set manufacturer, in consultation with the engine and generator manufacturers, will usually be in the best position to make the final decisions.

Reliability and Redundancy

Reliability

The reliability of a standby power supply will depend on a number of factors such as:

The design of the equipment that has been installed

The planning of the complete installation

The standard to which the equipment has been installed

The standard of maintenance

The operating procedures that are used including such matters as:
    Ensuring that fuel is available in the daily service tanks
    Ensuring that plant rooms are clean and not used as storage areas
    Personnel responding correctly to alarms initiated by the equipment
    Personnel following the operating and safety rules of the location

For high reliability each of the above activities must be undertaken to a high standard. An installation designed and planned to a high standard will be compromised if the operation or maintenance procedures are not of a similar standard.

Redundancy

Redundancy is introduced in Chap. 2, where it is given as one reason for the use of multiple sets. Redundancy may be defined as the provision of additional equipment such that, in the event of equipment failure, the remaining healthy equipment is capable of continuing to supply power to the load; manual intervention may be required.

A common example of redundancy is the ring main which is almost universally used for distribution at 11 kV. This allows, by manual intervention, the disconnection of a faulty section of cable and the restoration of supply to all consumers. If a second section of cable develops a fault, some consumers will be deprived of supply until a repair can be effected. An example of redundancy relevant to standby supplies is the provision of additional generating sets so that, in the event of a failure of one set, the redundant set can replace the failed set, either automatically or with manual intervention.

Redundancy for generating sets is expressed in two forms of notation, as a percentage or as the $N + 1$ notation. If the installation includes two sets each capable of supplying the entire load, the redundancy is said to be 100 percent or $1 + 1$; if there are three sets each capable of supplying one-half of the load the redundancy is said to be 50 percent or $2 + 1$, and so on. It is possible, although unusual, to have a redundancy above 100 percent. Thus, three sets each capable of supplying the entire load would have a redundancy of 200 percent or $1 + 2$.

The most common failure experienced by users of generating sets is a failure to start. Once having started, they usually continue running until stopped by their control systems or they fail due to some external circumstance such as a lack of fuel. For this reason, and because standby sets run only occasionally, reliability is usually assessed as the probability of a failure to start, rather than the mean time before failure which is applicable to continuously running equipment.

The Statistics of Redundancy

At the planning stage it may be necessary to calculate the increase in the probability of a successful start that can be achieved by providing redundant generating sets. The calculation is straightforward, the difficulty is at the beginning where it is necessary to assume a figure for the probability of a failure to start for a single engine. As stated at the start of this section, the reliability of a power supply is dependent on many factors, some of which require a good-quality crystal ball to assess and quantify.

Consider an installation of two equally sized sets (100 percent redundancy), each with a probability of a failure to start of 1 in 100, and a probability of a successful start of 99 in 100. On receipt of a start signal the following probabilities arise:

Event	Probability
Two sets start	$0.99^2 = 0.9801$
One set starts, one fails	$2 \times 0.99 \times 0.1 = 0.0198$
Two sets fail	$0.01^2 = 0.0001$

One of these events must result from the start signal and the arithmetic may be checked by addition: $0.9801 + 0.0198 + 0.0001 = 1.0$. The probability of a successful start is obtained by the addition of 0.9801 and $0.0198 = 0.9999$. These figures may be compared with the results for a single set which are:

Single set starts	0.99
Single set fails	0.01

It can be seen that the probability of total failure is reduced by 100 by the addition of the second set.

Most standby power installations use one or two generating sets, and the above calculations will be sufficient for most purposes. Where there are more than two sets the calculations follow a similar pattern and invoke the binomial theorem, which may be expressed as:

$$(a + b)^y = 1 \tag{3.3}$$

where a is the probability of a successful start for a single set
b is the probability of a failed start for a single set
y is the number of sets installed

The results of expanding this expression, using the probabilities of 0.99 and 0.01, for three, four, and five sets are:
For three sets:

$$0.99^3 + 3 \times 0.99^2 \times 0.01 + 3 \times 0.99 \times 0.01^2 + 0.01^3 = 1$$

For four sets:

$$0.99^4 + 4 \times 0.99^3 \times 0.01 + 6 \times 0.99^2 \times 0.01^2 + 4 \times 0.99 \times$$
$$.01^3 + 0.01^4 = 1$$

For five sets:

$$0.99^5 + 5 \times 0.99^4 \times 0.01 + 10 \times 0.99^3 \times 0.01^2 + 10 \times 0.99^2 \times .01^3 +$$
$$5 \times 0.99 \times 0.01^4 + 0.01^5 = 1$$

From these expressions, Table 3.2 may be constructed.

This table provides some insight into the working of this aspect of statistics. It is based on the assumed probabilities of 0.99 and 0.01 and readers will observe that the figures in column 2, where all sets start, are the various powers of the probability 0.99 and are independent of the probability 0.01. Similarly, the figures relating to all sets failing in columns 3–7, are independent of the probability 0.99 and are the various powers of the probability 0.01, which become insignificant in columns 5–7.

The figures in the other columns, where some sets start and some fail, are derived from the two probabilities and the binomial coefficients, which can be calculated somewhat laboriously but which are available from mathematical textbooks and from Pascal's triangle.

The probability of successful starting can be obtained from the table for any arrangement of redundancy. Most installations involving redundancy are arranged in the $N + 1$ form and success is achieved if all sets start or if one only fails. Thus, the probability of success is the sum of the figures in columns 2 and 3. For two-, three-, four-, and five-set

TABLE 3.2   Table of Probabilities

1 Number of sets	2 All start	3 1 fails	4 2 fail	5 3 fail	6 4 fail	7 5 fail	8 Check total
1 set	0.99000	0.01000	—	—	—	—	1.00
2 sets	0.98010	0.01980	0.00010	—	—	—	1.00
3 sets	0.97030	0.02940	0.00030	$<10^{-5}$	—	—	1.00
4 sets	0.96060	0.03880	0.00058	$<10^{-5}$	$<10^{-5}$	—	1.00
5 sets	0.95099	0.04803	0.00097	$<10^{-5}$	$<10^{-5}$	$<10^{-5}$	1.00

arrangements, the probabilities for success are found to be and 0.9999, 0.9997, 0.9994, and 0.99902, respectively. The corresponding probabilities for failure are the differences between the probabilities for success and unity, which are found to be 0.0001, 0.0003, 0.0006, and 0.00098, respectively. (These are also the sums of the figures for failure in columns 4, 5, 6, and 7.)

Routine Test Runs

It is most important that standby sets be regularly tested; despite this well-known fact many installations are either not tested or are tested in an unsatisfactory manner. Ideally, there should be a test run every 2 weeks, the period between tests may be extended but should not exceed a month.

The best test is to switch off the normal power supply and to allow the set (or sets) to run up to speed and supply the load as would be expected in a real mains failure situation. For many installations this is not allowed, sometimes the user does not have confidence in the standby power or in the uninterruptible power supplies due to problems or equipment failures in the past.

If real mains failure testing is not allowed, some other procedure has to be devised which proves the operation of the three subsequences, mains failure detection, engine starting sequence, and load acceptance. The engine control module may include a *Test Start* facility which proves the mains failure detection, if it does not it may be possible to prove operation by manual intervention such as removing a control fuse.

If testing on the real load is not allowed the installation should include a permanently installed test load which will provide the standby supply with a load not less than the minimum recommended by the engine manufacturer, probably 30 or 40 percent. This will prove all components in the sequence, including voltage regulation, speed regulation, and load sharing if applicable. This procedure introduces three small risks— mains failure detection may not be tested, any loading sequence is not tested, and the engine is not tested on full load. It may be that occasional testing can be arranged at weekends to prove all three of these functions.

Once started the set (or sets) should be left running until all temperatures have reached a steady state or for 1 h, whichever is longer. At the end of the test check the fuel remaining in the daily service tank, and if necessary and if it is not automatically done, top it up.

The test procedure described above assumes that the set will be attended while running. If there is a supply failure during a test run the control logic should ensure that the standby supply will supply the essential load, and in order to avoid an overload situation, the attendant must open the test load switch. If the set is not attended it will be necessary to include a contactor or circuit breaker electrically interlocked with the supply changeover device or standby supply circuit breaker.

For larger installations the advantages of a facility enabling the normal and standby supplies to be paralleled follow from the above. The set can be test started, paralleled, and loaded onto the real load before the normal supply is disconnected. It should be noted that this procedure does not test any loading sequence, a separate test should be devised and occasionally undertaken to prove this.

Kilowatts, Kilovars, and the Harmonic Components of the Load Current

Within an installation the standby supply will have a much higher impedance than the normal supply, and loads and load changes will have a much greater effect upon it.

The currents taken by the different loads may conveniently be considered as comprising four components in various proportions:

- The in-phase component of the fundamental frequency current, or power component, which on its own has a power factor of unity and represents kilowatts.

- The phase-quadrature component of fundamental frequency current, or reactive component, which on its own has a power factor of zero and represents kilovars.

- The harmonic or distorting currents which have no particular phase relationship with the fundamental current; they are parasitic and represent neither kilowatts nor kilovars.

- The notches in the fundamental sine wave of voltage caused by the delayed commutation of thyristor rectifiers.

For normal loads the kilowatt loading of a generating set will be the major component of the load, it is the component which determines the engine rating and will be larger than the kilovar or the harmonic loading. The application of kilowatt loading to an engine will cause a drop in speed which is restored to its correct value by the speed governor. The kilowatt loading is seen to a small extent by the generator, which experiences a slight drop in voltage due to the stator leakage reactance, but the resulting quadrature voltage drop has only a small effect on the terminal voltage which is restored to its correct value by the voltage regulator.

The kilovar loading of a generating set, conversely, has no effect whatever on the engine rating and will have no effect on the speed or governor setting. It will make a contribution to the kVA rating of the generator. The kilovar loading of the generator results in a rotor field demagnetizing effect, and a voltage drop across the stator leakage reactance, causing a drop in terminal voltage which is restored to its correct value by the voltage regulator. If the power factor of the load is lower than 0.8 the rotor field demagnetizing effect may be such that additional excitation is required which in turn may lead to a larger frame size of generator.

Harmonic currents may be regarded as being generated by the load and flowing through the generator of the standby set or through the supply system. Harmonic voltages are developed across the reactance of the generator (or of the supply system) which result in voltage distortion. Within the generator harmonic currents cause increased iron and copper losses. The voltage distortion affects current measuring instruments and introduces a distortion factor into calculations.

Notches in the fundamental sine wave of voltage are caused by the delayed commutation of thyristor rectifiers and the greater the delay the deeper the notches. Notching may affect the operation of voltage regulators and speed governors, and some form of filtering of the supplies to such devices may be necessary. The effect of notching is reduced if the generator includes a permanent magnet pilot exciter.

The size of the engine driving a generating set is determined only by the kilowatt loading, the size of the generator is determined by the kilovoltampere loading but may be increased due to the harmonic current loading. It follows that if a generating set is to supply a low power factor load or a heavily distorting load, the generator should be "oversized" with respect to the engine rating. An oversized generator has the advantage of having a lower impedance than one of standard rating and will be less affected by disturbances such as step loads, harmonic currents, and notches.

Characteristics of Particular Loads

The electrical characteristics of the load seen by a generator may have an effect on the quality of the output. In general where the load on the standby supply is a mix of various loads no problems will arise, but where one particular load predominates the likely effect should be considered at the planning stage.

Computer Loads

The dc power taken by computers is usually derived from switched mode power supplies which take a peaky current having typically a third harmonic component measuring 85 percent of the fundamental and many other harmonic currents producing an rms total harmonic distortion of 110 percent. In three-phase systems triplen harmonic currents have zero phase sequence and become additive in the neutral return path. In a distribution system supplying a large computer installation the neutral current can be large and can exceed the phase current; the system should be designed to allow for this. This statement applies whether the load is being supplied from the standby supply or from the normal supply.

Small computers may be connected directly to the supply, and while most computers of importance will be supplied from an uninterruptible power supply, they may on occasions be connected directly to the normal or the standby supply when the uninterruptible power supply is operating in bypass mode. Where standby power generation is at low voltage, the triplen harmonic current components will appear in the generator neutral and can cause overheating. If the total neutral current is likely to exceed say 15 percent of rated current, the generating set manufacturer should be advised.

Fluorescent and Discharge Lighting

These are nonlinear loads taking a current rich in odd harmonics. BS EN/IEC 61000-3-2 allows a third harmonic of up to 30 percent. The same problems arise with neutral currents as are described for computer loads in the preceding subsection.

Cyclically Varying Loads

A load such as a rotating radar aerial will vary cyclically at a regular rate due to the aerial's rotation and the more or less static direction of the wind. If the regular rate of variation is comparable to the natural frequency of response of the engine governor, continuous hunting about the mean speed may occur. Continuous motion of the governor parts and linkages can lead to premature wear and to lost motion in the fuel control mechanism. Increasing the speed of response of the governor may lead to instability, and reducing the speed of response may result in the frequency not remaining within specification. The generator set manufacturer should be aware of a significant load of this nature at an early stage of the contract.

Motors

Any large motors in the load will affect the supply voltage during starting. An induction motor will take a current of six or seven times its rated current at a power factor of say 0.15 when started direct on line. A squirrel cage motor taking a starting current equal to the generator rated current may cause a drop in voltage, depending on the type of excitation, of up to 20 percent. During the accelerating period the current reduces but the power factor increases, and the peak starting power taken may well be of the order of 200 percent of rated power for a very short period. A large motor will have a run-up time longer than the voltage recovery time and once the voltage regulator has restored the initial dip it will take care of the comparatively slow changes in current and power factor that occur during the accelerating period.

More is written about the effect of motor starting in an earlier section titled "Sizing the Engine and Generator."

Contactors associated with motor control gear can experience difficulties. If motor starting causes the voltage to drop momentarily to below say 80 percent, there is a danger that the contactor will fail to close its magnetic circuit, with disastrous consequences for the contacts and the operating coil.

If a generator that is already supplying some running motors experiences a step load, the resulting voltage dip will be increased by the running motors. This is because ac induction motors run at almost constant speed and therefore behave as constant power machines; if there is a drop in the supply voltage there must be a corresponding rise in current which increases the step loading. The effect is contrary to that produced by resistive and most other loads, which take less current during voltage dips. The effect will be noticeable only if the motor loading is significant, say in excess of 50 percent of the generator rating.

Power Factor Correction Capacitors

If an installation includes bulk power factor correction equipment, consideration should be given to disconnecting it when the standby supply is in use. The essential load will be less than the total load and there may be overcorrection unless the capacitor bank can be reduced. A generator supplying a capacitive load is likely to produce overvoltages, certainly the voltage regulator will have less control. When the standby supply is in use there are no tariff implications of a low power factor, and bulk correction seems to have no benefit except to reduce the generator current.

Regenerative Loads

The regenerative load most likely to be encountered in buildings is a lift. A passenger elevator will usually feed power back into the supply system when raising an empty car or lowering a full car. When the normal supply is in use, surplus power can be absorbed by other building loads, or be fed back into the supply system, but when the standby supply is in use different conditions apply. The elevator manufacturer can advise the maximum power that will be regenerated (assuming no overloading), and the generator set maker can advise the power required to drive the set at rated speed. Any surplus power must be absorbed by other building loads or by a dummy load provided for the purpose. Under regenerative conditions the governor will already be on minimum fuel and can have no effect. The power absorbed by generating sets depends on speed and size, but varies between 7.5 and 25 percent of the kW rating.

Step Loads

Any step load applied to a generating set will produce dips of both voltage and frequency. The voltage dip will be approximately proportional to the reactive kVA component; full load kVA at 0.8 power factor will result in a drop in voltage, depending on the type of excitation, of up to 16 percent. At lower loads it will be proportionally less, thus at 50 percent load a drop of up to 8 percent might be expected. The frequency dip will be proportional to the kW component and the maximum step load permitted by the engine will probably result in a speed dip of say 8 percent. Both effects are transient, the voltage regulator and the speed governor will restore voltage and frequency to within tolerance within a few seconds.

Uninterruptible Power Supplies

Any uninterruptible power supplies within the installation will form part of the essential load. The input modules of many types of power supplies are rectifiers which take a current rich in harmonics; theoretically the current taken by an idealized three-phase bridge rectifier has an rms harmonic content equal to 30 percent of the fundamental, which can be reduced to 14 percent by using two phase shifted bridges. The most troublesome harmonic currents are the fifth and seventh or, for the phase shifted bridges, the eleventh and thirteenth, there are no triplen or even harmonics. The harmonic currents cause additional heating in the windings and in the rotor but, provided that the power supply loading on the generator does not exceed say, 50 percent of the generator rating, difficulties are unlikely to arise.

There are two types of rectifier in use, diode and thyristor and the preceding comments apply to both; diode rectifiers use natural commutation at the voltage changeover point, and thyristor rectifiers use delayed commutation and can cause severe notching of the generator output voltage. Notching may affect the operation of voltage regulators and speed governors, and some form of filtering of the supplies to such devices may be necessary. The effect of notching is reduced if the generator includes a permanent magnet pilot exciter. Notching is discussed in Chap. 4.

Vibration

A detailed description of the effects of vibration is beyond the scope of this book, advice should be sought from the engine maker or other specialist. However, this section may be helpful and is intended only as an introduction to vibration control. It is concerned with the vibration of the set in the vertical direction, but the set has other degrees of movement such as, to use a nautical analogy, rolling, pitching and yawing. For example, during an electrical fault or at load application, inertial energy is extracted from the rotating parts and the resulting forces cause a rolling motion of the set. Usually, only the engine maker has sufficient information to take into account these more complex matters.

Diesel Engine–Driven Generating Sets on Solid Foundations

A diesel engine has reciprocating parts which introduce out of balance forces leading to vibration. In addition, there will be minor out of balance forces due to dynamic unbalance of the generator or engine rotating parts. The vibration cannot be prevented and has to be accommodated by so-called antivibration mountings which allow the whole generating set to move, under controlled conditions and in a vertical direction, relative to the floor which supports it.

Within a building vibration must be kept within tolerable limits, and if the building is occupied the need for vibration control is greater. For sets having the generator solidly spigot-mounted on the engine body, antivibration mountings are usually installed between the engine/generator mass and an underbase which rests on a solid floor. For sets where the engine and generator are connected through a flexible coupling, the two masses are usually solidly mounted on a rigid base frame which rests on antivibration mountings resting on the specially prepared floor pads. This arrangement is used only for larger sets and an additional underbase is not normally provided.

The mountings may use coil springs in compression or rubber in shear. Coil springs have a single linear characteristic but rubber is a material which increases in stiffness when subject to vibration, its dynamic and static characteristics therefore differ. When selecting rubber mounts it is important to use the dynamic characteristics. Two pairs of mountings, a pair on each side of the set, is the minimum number and if the disposition is symmetrical about the center of gravity the loading on each will be equal. If more than four mountings are used, attention must be given to the weight distribution.

From the vibration viewpoint, the best location for the generating set is on a rigid solid floor at basement or ground level, the choice of mountings then requires three basic rules to be followed:

1. To avoid resonance each of the antivibration mounts and its share of the supported mass must have a resonant frequency well below, say one third of, the lowest forcing frequency which the engine produces.

2. To avoid rocking or pitching it is essential that the static deflection of each mount on the generating set is the same. If the weight distribution is unequal this will require the selection of different mountings.

3. The antivibration mounts must be properly designed and capable of supporting the applied load.

   Provided that the weight resting on each mount and the corresponding deflections are known, compliance with the first rule is fairly simple. The lowest forcing frequency should be ascertained from the engine or set manufacturer, it will be associated with the speed and number of cylinders. The natural frequency of vibration of the mass on its mounting is dependent on the static deflection for a steel spring or upon the dynamic deflection for rubber mountings in accordance with the following formula:

$$f_o = \frac{15.8}{\sqrt{\delta}} \text{ Hz} \qquad (3.4)$$

where f_o = the natural frequency
      δ = the deflection in mm

   If the lowest forcing frequency is advised as 25 Hz, the natural frequency of the mountings with their supported mass should not exceed 8 Hz and the deflection (static for springs, dynamic for rubber) should not be less than 4 mm.

   Antivibration mounts allow the set a freedom of movement and it is important that all services to the set pass through flexible sections. The services include electrical cables, fuel, oil and water pipes, ductwork, and exhaust pipes.

   The foregoing treatment covers basic concepts only and does not, for example, consider transmissibility of forces to the floor, or the effect of damping. Steel springs will be lightly damped but rubber mounts more so. In general, specially damped mounts are not necessary for generating sets.

Diesel Engine–Driven Generating Sets on Suspended Floors

If the set is mounted on a suspended floor within a building, the selection of the mountings has to take into account the resilience of the floor which will itself tend to act as one large antivibration mounting.

All that is written in the previous subsection is relevant but in addition it is necessary to consider the natural frequency of vibration of the floor, which will almost certainly be lower than the lowest forcing frequency which the engine produces. To avoid resonance each of the antivibration mounts and its share of the supported mass must have a resonant frequency well below, say one half of, the natural frequency of the floor. It will be difficult to ascertain the natural frequency of the floor and guidance from a structural engineer may be needed. If the deflection of the floor due to the load of the set can be ascertained, the natural frequency can be calculated using the formula already given. It will be found that the deflection of the mounts should be at least four times the deflection of the floor.

To achieve the above, the antivibration mountings may require a static deflection of up to 50 mm to avoid any possibility of resonance and to isolate the movement of the set from the building structure. Very large forces can be involved during starting, during fault clearance and during faulty synchronizing, and the freedom of movement can lead to instability at such times, particularly if the center of gravity of the set is high. To overcome this problem expensive remedies are sometimes necessary such as the addition of an inertia block (below floor level) to lower the center of gravity of the set.

When the antivibration mounts allow a large freedom of movement it is particularly important that all services to the set pass through flexible sections. The services include electrical cables, fuel, oil and water pipes, ductwork, and exhaust pipes.

The location and arrangement of the generating set should always be agreed with the structural engineer responsible for the building.

Gas Turbine–Driven Generating Sets

A gas turbine has no reciprocating parts and the only out-of-balance forces that arise are those due to the dynamic unbalance of the generator, turbine, or gearbox rotating parts. The problem is very much simpler and the use of the correct antivibration mountings is usually adequate to ensure smooth running. All services must pass through flexible sections as described for diesel engines.

The location and arrangement of the generating set should always be agreed with the structural engineer responsible for the building.

Noise

The Units Used in Noise Measurement

The human ear is sensitive to a very wide range of sound intensities. The lowest sound intensity that the normal ear can detect is 10^{-12} watts per m², and the highest without pain is about 10 watts per m², a power ratio of 10^{13}. If used on a linear scale this range would be unmanageable and a logarithmic scale is used such that:

Sound power level (SWL)

$$= 10 \log_{10} \frac{\text{Sound power (watts/m}^2)}{10^{-12} \text{ watts/m}^2} \text{ decibels (dB)} \qquad (3.5)$$

The ear is in fact responsive to sound pressure levels and in measuring these the same reference level is used and 10^{-12} watts/m² becomes 2×10^{-5} newtons/m². Sound power is proportional to the square of the pressure and since decibels represent a power ratio the corresponding equation becomes:

Sound pressure level (SPL)

$$= 20 \log_{10} \frac{\text{Sound pressure (newtons/m}^2)}{2 \times 10^{-5} \text{ newtons/m}^2} \text{ decibels (dB)} \qquad (3.6)$$

The response of the ear to changes of intensity is more logarithmic than linear so the decibel is a better indication of what we hear than is a linear scale of pressure. Furthermore, the decibel is said to be the smallest change in intensity that the normal ear can detect, although the statement is subjective and frequency dependent.

The Sound Frequency Spectrum`

The range of sound frequencies of interest in connection with diesel engines and gas turbines is from 22 to 11,000 Hz. In order to make it possible to consider and to make calculations of acoustic performance, it is customary to divide the spectrum into nine octave bands, defined by a geometric center frequency as in Table 3.3.

TABLE 3.3 Octave Frequency Bands (Hz)

Center frequency	Lower limit of band	Upper limit of band
31.5	22	44
63	44	88
125	88	176
250	176	353
500	353	707
1000	707	1,414
2000	1414	2,825
4000	2825	5,650
8000	5650	11,300

If a more precise measurement is required each octave band is divided into three bands of one-third of an octave, thus the highest band would be divided into three bands with center frequencies of 6300, 8000, and 10,000 Hz.

In order to establish the acoustic performance of a machine, manufacturers measure sound pressure levels for each of the octave bands at a distance of say, 1 m, from the machine, at a number of positions around the periphery. From these pressure levels it is possible to calculate by a process of integration the total sound power emitted in each of the octave bands. These data are essential and are the starting point for any acoustic calculations that have to be performed.

The human ear is not equally sensitive to all frequencies, in fact it is least sensitive to low frequencies and most sensitive to midrange frequencies. A sound level meter accepts all the frequency components of the sound and adds all their levels together to indicate an overall sound intensity. It is more useful to use a subjective indication which takes into account the frequency response of the ear, and this is achieved by incorporating in the meter a weighting network which attenuates the low and high frequencies in accordance with a defined curve. There are three such standard curves known as A, B, and C weighting curves and the one of interest in this context is curve A which, it is generally agreed, represents the sensitivity of the human ear to the range of audible frequencies. All measurements which have been weighted in accordance with the curve are expressed as dBA and it is important that the distinction between dB and dBA is recorded.

Diesel Engines and Gas Turbines as Noise Sources

A typical noise spectrum of a diesel engine is given in Table 3.4 which shows dBA sound power figures calculated from measurements on a 1.2 MW, 1500 rpm, 16-cylinder V engine running on full load, part load, and no load.

A typical noise spectrum of a gas turbine is given in Table 3.5 which shows unsilenced dB sound power figures calculated from measurements on a 2.5 MW class turbine running on full load. A gas turbine

TABLE 3.4   Frequency Spectrum for a 1.2 MW Diesel Engine

Center frequency Hz	125	250	500	1000	2000	4000	8000
1200 kW dBA SWL	107	109	113	111	110	107	107
1000 kW dBA SWL	105	108	111	109	109	105	102
No-load dBA SWL	103	107	110	109	109	104	94

TABLE 3.5   Frequency Spectrum for an Unsilenced 2.5 MW Class Gas Turbine

Center frequency Hz	63	125	250	500	1000	2000	4000	8000
Sound power level					dBA			
Air inlet dB SWL	110	113	116	119	121	124	124	133
Exhaust dB SWL	130	134	135	135	132	130	127	124
Casing dB SWL	98	102	106	110	112	112	112	111

TABLE 3.6   Frequency Spectrum for a Sound Attenuated Packaged 2.5 MW Class Gas Turbine

Center frequency Hz	63	125	250	500	1000	2000	4000	8000	
Sound pressure level					dB				dBA
Machine acoustic enclosure	88	85	82	76	74	71	69	64	80
Turbine air inlet	96	90	74	65	65	64	69	76	80
Turbine exhaust	101	90	79	73	70	68	63	64	80
Enclosure ventilation exhaust	91	86	80	65	61	60	68	78	80
Oil cooler exhaust	81	81	88	69	65	64	60	62	80

package will normally include sound attenuation on all these sources of noise and typical attenuated sound pressure levels 1 m from the source are given in Table 3.6.

The Noise Level within the Engine Room

The sound produced from the machine travels to the room boundaries and is reflected several times from the walls, ceiling, and floor and at each reflection there is some absorption of sound. The reverberation time of a room can be calculated and is the time taken for a sound to fall through 60 dB and depends on the size, the contents, and the materials used in the construction of the room. The reverberation time will be different for each frequency band. Knowing the acoustic properties of the room and the sound power released within it, the resulting sound pressure level (SPL) within the room can be calculated. The sound pressure level within an engine room will be high and may exceed 110 dB, it can be reduced by introducing sound absorbent material within the room or by providing an acoustic hood over the machine.

Acoustic Treatment of Ventilation Openings

If acoustic attenuation is required at ventilation openings, and it almost certainly will be, it can be achieved by the installation of acoustic louvers or of splitter units. There will be an air pressure drop across the louver or splitter dependent upon the face velocity, and supply (or extract) fans must be adequate for the duty.

Acoustic louvers are of the order of 300 mm deep and provide moderate attenuation, typical performance is indicated by Table 3.7. Acoustic louvers are normally weatherproof but cannot be closed so additional pivoted weatherproof louvers will be required if the room is to be heated in the winter.

Splitter units comprise a length of duct up to 2.5 m in length which, as the name implies, is split into several vertical sections by barriers of sound absorbent material. There is scope for design variation in the length of the unit and the width of the vertical sections, and manufacturers offer a wide range of standard units; typical performance is indicated by Table 3.8. Acoustic performance can be improved by narrowing the width of the airway sections but at the expense of increasing the air pressure drop. The acoustic performance of splitter units is slightly better for air inlet units than for air outlet units; for air inlets the air and sound are travelling in opposite directions. Splitter units are not provided with any weather protection and weatherproof louvres will be required at the exposed end.

The Noise Level outside of the Engine Room

There will be noise breakout from the engine room to the outside and this will be directly dependent upon the internal noise level. There will be noise from the exhaust outlet and a small amount from the surface of the exhaust pipe itself. If there is external machinery such as a remote radiator and fan this will be a source of additional noise. The noise breakout will be through the engine room walls and ceiling and through any ventilating and cooling air inlet and outlet openings, or any other services that pass through the wall. The breakout through the walls and ceiling themselves depends on the sound reduction index

TABLE 3.7   Typical Performance for Acoustic Louvers

Center frequency Hz	63	125	250	500	1000	2000	4000	8000
Attenuation dB	4	4	6	11	17	20	15	15

TABLE 3.8   Typical Performance for a 2.5 m Splitter Unit

Center frequency Hz	63	125	250	500	1000	2000	4000	8000
Attenuation dB	9	19	30	40	40	21	15	11

of the materials of which they are constructed, in general the heavier the material the greater the reduction.

Knowing the sound pressure level within the room, the sound reduction index of the materials used in its construction, and the acoustic properties of any louvers or other sound attenuating items, it is possible to calculate the sound pressure level at defined positions outside of the room. The assessment of the resulting sound pressure levels at remote locations takes into account the distance (the inverse square law applies), and for distances greater than a few hundred meters, molecular absorption adds to the attenuation. There is likely to be additional attenuation due to screens and barriers such as walls, but it is safer to regard this as a bonus rather than take it into account.

Local Authority Requirements

In the United Kingdom the local authority will have its own rules for noise limitation, and in mixed residential and industrial areas these will have been devised to minimize complaints from local residents. Standby generating plant runs only occasionally and is less likely to cause complaints than continuously running plant. A method for rating industrial noise in mixed areas is described in British Standard 4142. The method is to compare the new noise level at specific sites with the existing background noise level over a 24-h period, and three guidelines are given:

1. If the new source is 10 dB above the existing noise, complaints are likely.
2. If the new source is 5 dB above the existing noise, it is of marginal significance.
3. If the new source is more than 10 dB below the existing noise, complaints are unlikely.

The British standard also includes the warning that noise assessment is a skilled operation and should be undertaken only by persons who are competent in the procedures.

Safe Working Procedures

Access to the generating set should be restricted to personnel who are authorized to work on or near it. If the set is located within an engine room, the room should be locked unless access has been authorized and a key issued to allow maintenance, repair, or cleaning. If the set is located in an open area it should be protected by lockable doors or removable covers.

Unauthorized access should be prohibited to prevent persons tampering with engine controls or adjustments and to avoid the dangers that arise while it is running or when it unexpectedly starts. There should be a rule that before any work on the set is commenced, the starting sequence should be inhibited. Anyone who has experienced an unexpected start while manually topping up a baseframe service tank will realize the danger that can arise due to the sudden noise and engine movement, and the involuntary reactions that occur.

Before any major work is undertaken on the generating set, the generator isolator or circuit breaker should be locked open, the key being in the care of the senior person working on the set. This prevents any possibility of the generator being connected to a mains supply, which would cause it and the engine crankshaft to rotate. The overcurrent protection may operate but too late to avoid any danger.

Large organizations usually have a set of safety rules including a permit to work system, and if standby generation is added to a particular installation some additional rules may be necessary. Smaller organizations frequently get by with the minimum of safety rules. In such cases it is recommended that some relevant rules be drafted and clearly displayed within the engine room, or near the set if it is in an open area. There should be a notice on the access door or doors making it clear that access is restricted to persons who have been authorized to work on the set or in the room. There should also be prominently displayed a danger notice advising that the set starts automatically.

If the set is controlled from a remote station it is important that a set of safety rules is agreed between the local and remote stations. Before any start is initiated the rules must ensure that no person is working on or near the set.

Within the engine room there should be emergency lighting which, on loss of mains, provides illumination until power is restored. This is sometimes provided by removable battery-operated hand lamps which plug into wall-mounted base units. On loss of mains the lamps automatically provide static illumination, and if removed from the base units may be used as portable hand lamps. After use the lamp is returned to the base unit where the battery is automatically recharged.

Bibliography

International Standards

ISO 3046, BS 5514—Reciprocating internal combustion engines. Performance. (For details, see Chap. 1 Bibliography.)

ISO 8528, BS 7698—Reciprocating internal combustion engine driven alternating current generating sets. (For details, see Chap. 1 Bibliography.)

British and European Standards

BS 4142—Method for rating industrial noise affecting mixed residential and industrial areas

BS EN IEC 61000—Electromagnetic compatibility

Part 3-2 Limits for harmonic current emissions (equipment input current up to 16 A per phase)

Other Documents

Sharland, I., *Woods Practical Guide to Noise Control*, Woods of Colchester Ltd., Colchester, Essex, United Kingdom.

Chapter

4

Harmonic Distortion of the Supply

Acknowledgment

This text originates from Chap. 10 of a book titled *Uninterruptible Power Supplies* published in 1992 by Peter Peregrinus Ltd. on behalf of the Institution of Electrical Engineers of London. It has been modified in parts but much of the material is unchanged and is published here with the permission of the Institution of Electrical Engineers.

Nonlinear Loads and Current Distortion

Introduction

A nonlinear load may be defined as a load which, having a sinusoidal voltage applied to it, passes a nonsinusoidal current. Many everyday loads are nonlinear but the nonlinearity is often unimportant; an unloaded transformer is nonlinear owing to magnetic saturation. Significant nonlinear loads associated with UPS equipment include rectifiers and switched mode power supplies.

Any regular nonsinusoidal waveform may be regarded as a compound wave made up from a fundamental component and harmonic components; the composition of such waveforms may be established mathematically by applying Fourier's analysis. Even harmonic currents and voltages indicate a lack of symmetry between positive and negative half cycles.

The phase sequence of three-phase harmonics is in accordance with the following pattern:

Harmonic number	1	2	3	4	5	6	7	8	9	10	11	12
Phase sequence	+	−	0	+	−	0	+	−	0	+	−	0

The concept of harmonic phase sequence may not be familiar to readers and Table 4.1 may offer an explanation.

It will be observed that the harmonic orders 3, 6, 9, 12, etc. have zero phase sequence, they are divisible by three and are known as triplen harmonics or triplens. They are of particular importance because in a three-phase system the phase currents become additive in the neutral. In a balanced system the triplen current flow in the neutral will be three times what is in each phase conductor. If there is no neutral conductor (a three-wire system) three-phase triplen harmonic currents cannot flow.

The rectifiers normally included in UPS equipment are of the three-phase bridge type; as there is no neutral connection there will be no significant triplen harmonic currents and as positive and negative half cycles should be balanced there will be no significant even harmonic currents. Any that are present will be due to a lack of symmetry between the firing of pairs of thyristors. The half-controlled bridge generates even harmonics but is not usually found in UPS equipment and will not be discussed further.

The "power factor" of a nonlinear load can have several meanings; whenever the term is used it should be defined. Assuming a pure sine wave of voltage, only the fundamental current component can supply power, any harmonic components of current are parasitic. Industrial ammeters may indicate the rms or the rectified average value of either the fundamental or of the total current, and it is not always clear what is being indicated. To calculate power from current and voltage readings it is necessary to include a distortion factor μ in addition to the conventional power factor, $\cos \phi$ of the fundamental current and voltage components.

TABLE 4.1   Explanation of Harmonic Phase Sequence

Phase	A	B	C	
	Phase displacement, degrees			Sequence
Fundamental	0	120	240	A B C
5th	0	$5 \times 120 = 600$ $= 240$	$5 \times 240 = 1200$ $= 120$	A C B
6th	0	$6 \times 120 = 720$ $= 0$	$6 \times 240 = 1440$ $= 0$	Zero
7th	0	$7 \times 120 = 840$ $= 120$	$7 \times 240 = 1680$ $= 240$	A B C
8th	0	$8 \times 120 = 960$ $= 240$	$8 \times 240 = 1920$ $= 120$	A C B

Conventionally, nonlinear loads are regarded as loads which take current at the fundamental frequency and which include, in series, harmonic current generators having zero impedance.

There are a number of formulae likely to be encountered in connection with nonlinear loads; they appear below and use the following convention:

a = rms value of the total current

a_1 = rms value of the fundamental current component

a_n = rms value of the nth harmonic current component

The rms value of total current = a

$$= \sqrt{a_1^2 + a_2^2 + a_3^2 + a_4^2 \cdots} \qquad (4.1)$$

Distortion factor μ = $\dfrac{\text{rms value of the fundamental current component}}{\text{rms value of the total current}}$

$$= \frac{a_1}{\sqrt{a_1^2 + a_2^2 + a_3^2 + a_4^2 \cdots}} \qquad (4.2)$$

Total harmonic distortion = $\dfrac{\text{rms value of the harmonic components}}{\text{rms value of the fundamental component}}$

$$= \frac{\sqrt{a_1^2 + a_2^2 + a_3^2 + a_4^2 \cdots}}{a_1} \qquad (4.3)$$

Total harmonic distortion is often abbreviated to thd and is usually expressed as a percentage of the fundamental component (either current or voltage).

Peak or crest factor = $\dfrac{\text{instantaneous peak value of current}}{\text{rms value of the total current}}$

$$= \frac{\text{instantaneous peak value of current}}{\sqrt{a_1^2 + a_2^2 + a_3^2 + a_4^2 \cdots}} \qquad (4.4)$$

For a sinusoidal supply voltage, and where $\cos \phi$ is the power factor of the fundamental current component:

$$\text{Power} = v\, a_1 \cos \phi$$

$$= v\, a\, \mu \cos \phi \qquad (4.5)$$

The power factor, cos φ, of a diode rectifier is close to unity whereas that of a thyristor rectifier is determined by the commutation delay angle such that

$$\cos \phi = \frac{\text{output dc voltage with phase control}}{\text{maximum dc voltage with no delay}} \tag{4.6}$$

which can result in unexpectedly low power factors for lightly loaded thyristor rectifiers. Typical values for μ and cos φ are:

	μ	cos φ
Diode rectifier	0.96	0.98
Thyristor rectifier	0.94	0.5–0.9

Wave Shape or Profile

The profile of a nonsinusoidal quantity is determined by the magnitudes and phase relationships of its harmonic components. It cannot be determined from a knowledge of the harmonic magnitudes alone, phase relationships are involved; hence the use of the peak or crest factor. The currents indicated by Figs. 4.1a and b both have the harmonic components (from fifth to thirty-first) characteristic of a six-pulse rectifier. For Fig. 4.1b the signs or "polarities" of the fifth, eleventh, seventeenth, twenty-third, and twenty-ninth harmonics have been reversed to achieve the high current peak. For Fig. 4.1a the addition of the infinite number of higher characteristic harmonics would result in a waveform having vertical edges and a flat top.

The rms magnitudes of the harmonics of both waveforms are:

n	1	5	7	11	13	17	19	23	25	29	31
rms	0.78	0.156	0.11	0.07	0.06	0.046	0.04	0.034	0.031	0.027	0.025

Data relating to the two waveforms may be calculated:

Harmonics Generated by Bridge Rectifiers

Rectifier Operation

Many UPS assemblies take power from the mains to feed a rectifier input module. A common rectifier configuration is the three-phase bridge, which theoretically demands a current with a harmonic content 30 percent of the fundamental. These harmonic currents, in flowing through the impedances of the supply system, cause distortion of the supply voltage for local consumers.

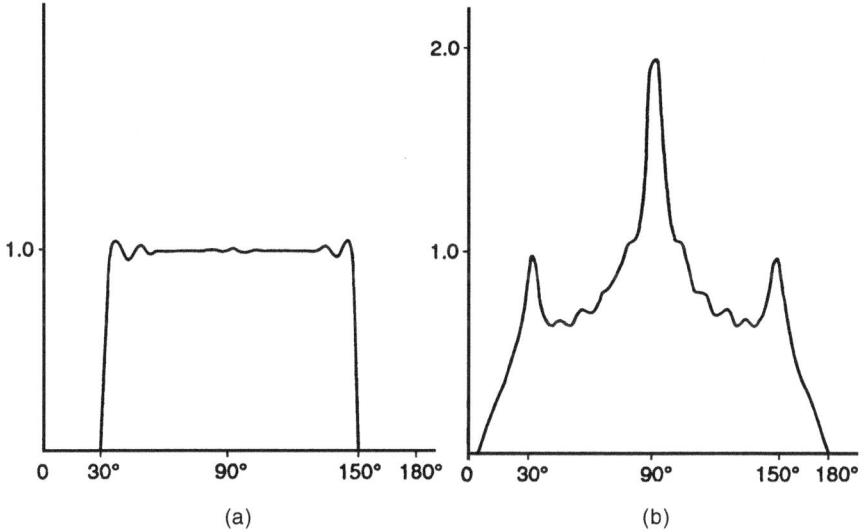

Figure 4.1 Two waveforms with the same harmonic content. (*a*) A quasi-square wave with harmonics up to the thirty-first. (*b*) The fifth, eleventh, seventeenth, twenty-third, and twenty-ninth harmonics are reversed in phase.

TABLE 4.2   Data Relating to Fig. 4.1

	Quasi square wave (4.1*a*)	Peaky wave (4.1*b*)
Fundamental rms a_1	0.780	0.780
Total rms a	0.813	0.813
Distortion factor μ	0.959	0.959
Total harmonic distortion	0.294	0.294
Peak or crest factor	1.244	2.404

A three-phase bridge rectifier operates as two complementary star groups, each arm conducting for 120° in each cycle and the two groups being displaced by 180°. The operating sequence of the arms and the idealized current waveforms are indicated in Fig. 4.2.

It can be shown from a Fourier analysis that the harmonic content of such quasi square waveforms includes only the harmonic numbers derived from the formula:

$$n = px \pm 1 \tag{4.7}$$

where n = the harmonic number
  p = the number of current pulses per cycle (6 for a three-phase bridge)
  x = any integer

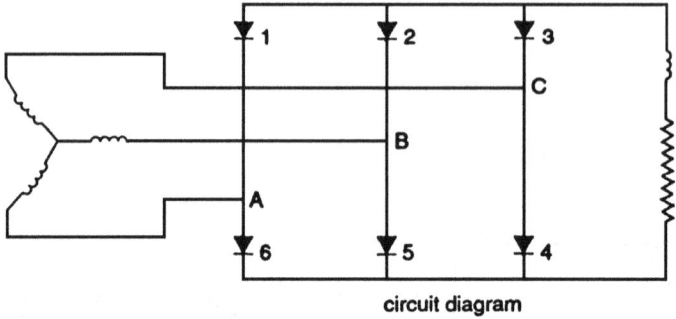

circuit diagram

diode conduction sequence

idealised currents

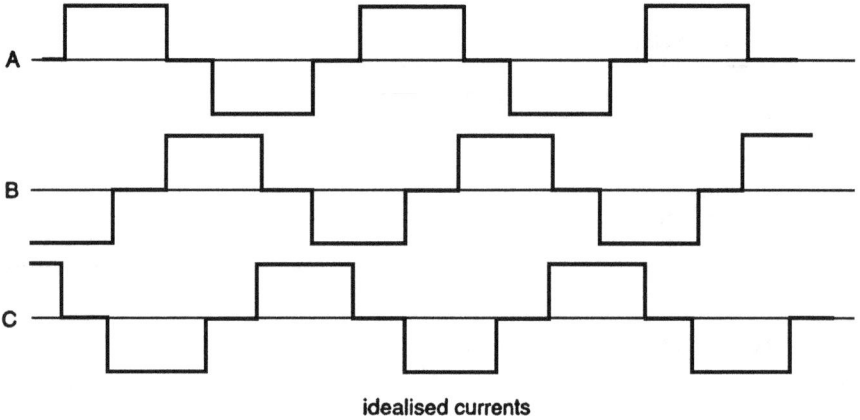

Figure 4.2 Circuit diagram, operating sequence, and idealized currents of a three-phase bridge rectifier.

It follows that a three-phase bridge will take harmonic currents of the orders 5 and 7, 11 and 13, 17 and 19, etc. and if the pulse number is increased to 12 by adding a second bridge with a phase displacement of 30°, harmonic currents of the orders 5 and 7, 17 and 19, etc., will be suppressed leaving only the orders 11 and 13, 23 and 25, etc. The Fourier analysis indicates that the magnitudes of these harmonic currents are similarly based on a simple mathematical relationship.

Whatever the pulse number the magnitude of the nth harmonic is inversely proportional to the harmonic number, thus the fifth harmonic will be 20 percent of the fundamental and the thirteenth harmonic will be 7.7 percent. The total harmonic content of the current for a three-phase bridge (six-pulse) is 30 percent of the fundamental and for a double bridge (12-pulse) 13.9 percent.

Commutation

The rectangular waveforms of Fig. 4.2 would require an instantaneous transfer of current from the conducting phase to the next in sequence. Such instantaneous transfer cannot occur owing to the presence of inductance in the supply; in practical circuits commutation between diodes requires a time measured in milliseconds. Commutation has the effect of slightly reducing the higher harmonic content of the line current.

During commutation the direct current is supplied from two diodes connected to different transformer phases. As both diodes are conducting and connected to a common point the transformer phases may be considered as being connected together. The common point assumes a voltage that is the mean of the two phases until commutation is complete. Figure 4.3 illustrates the effect of diode commutation on the supply voltage and on the current output.

With a diode rectifier commutation occurs at practically zero voltage difference, and rates of change of currents are not high. If the rectifier is of the controlled type it will incorporate thyristors instead of diodes and reduction of the dc output voltage will be achieved by delaying the commutation. During delayed commutation the voltage difference between the phases can be high, which leads to a high rate

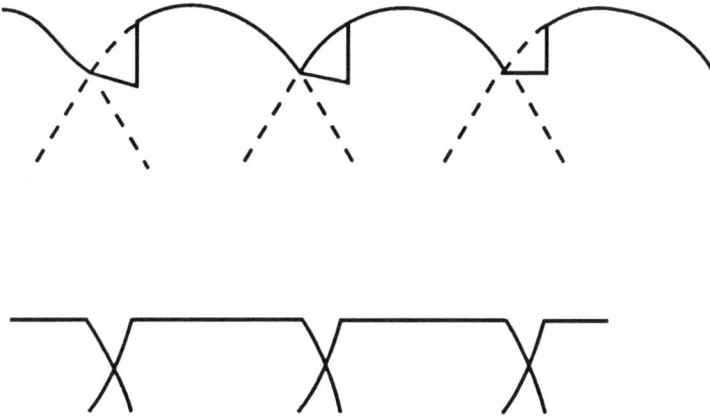

Figure 4.3   Effect of diode commutation on voltage and current.

of change of current and a short commutation period. The idealized current shapes are not affected, but in practice, since the load inductance is finite, delaying the commutation leads to a change of current shape.

As in the case of diodes, during commutation the voltage of the common point assumes a voltage that is the mean of the two phases, leading to a voltage rise in the preceding and a dip in the succeeding phase. When commutation is complete the previously conducting thyristor abruptly assumes a reversed bias state. If the supply inductance is small it may be necessary to add series inductance to limit the commutating di/dt. Figure 4.4 illustrates the effect of delayed commutation on the supply voltage and on the current output.

The Effect of a Bridge Rectifier on a Supply System

Distortion of the Supply Voltage

A rectifier may be regarded as a load which takes power from the supply system at fundamental frequency and generates harmonic currents which are fed back into the system. In flowing through the system the harmonic currents cause voltage drops along the various current paths, and these voltages distort the system voltage waveform. The distortion will be greatest at the rectifier terminals, and the degree of penetration into the system will depend on the system impedance. If a number of rectifiers feed harmonic currents into different points of a system the effect is cumulative, the penetration can be deep, and a large number of consumers may be affected.

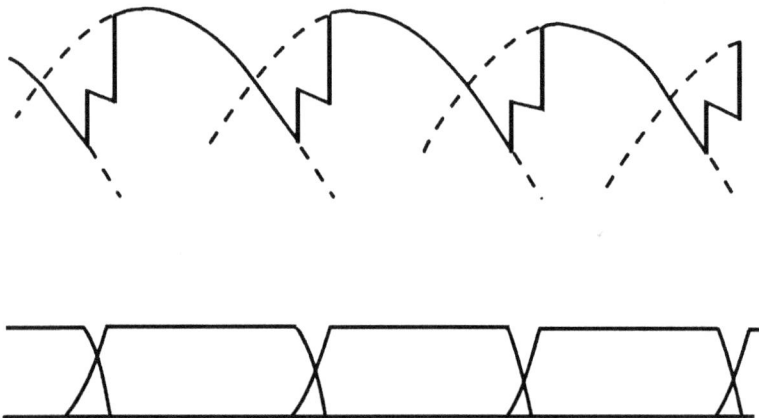

Figure 4.4   Effect of delayed commutation on voltage and current.

A calculation of the effect of harmonic currents on a supply system is complex and requires a knowledge of the resistance and reactance of the various components such as overhead lines, cables, and transformers. If it is assumed that the supply impedance is purely inductive the voltage distortion is related to the formula:

$$\frac{\text{Harmonic voltage}}{\text{System voltage}} = \frac{n \times I_N}{I_{SC}} \qquad (4.8)$$

where n = harmonic order
I_N = harmonic current of order n
I_{SC} = symmetrical prospective fault current

This formula does not provide a reliable prediction of the voltage distortion and Engineering Recommendation G.5/4 (see below) includes factors to modify the harmonic impedance to allow for cable capacitance, overhead line inductance, and other random effects.

Any resistive loads connected to the system provide additional shunt paths for harmonic currents and reduce the voltage distortion. Harmonic currents and the resulting voltage distortion are undesirable for several reasons, the circulating currents cause additional system losses, the distortion can affect the operation of equipment connected to the system, the losses and temperature rises of motors and other machines are increased, and power factor capacitors can be damaged.

Some Effects of Distortion

Data and control circuits running parallel to circuits carrying distorted currents may be affected by inductive coupling particularly if the severe notching associated with thyristors is present. Telephone circuits may similarly be affected; the human ear is particularly sensitive to frequencies of about 800 Hz, the sixteenth harmonic of 50 Hz. A three-phase bridge rectifier produces the seventeenth harmonic.

Within asynchronous induction motors harmonic stator currents cause flux systems which rotate at harmonic speeds and cause additional iron losses, particularly in the rotor. The effects of harmonic currents within synchronous machines is discussed in the section titled "The Effect of Bridge Rectifier Loads on Local Generators."

Analogue or digital measuring instruments may display a true rms value or a rectified average multiplied by the sine wave form factor (1.11). If the wave form is not sinusoidal the form factor will differ from 1.11 and the reading will be incorrect. Even more confusion is caused if the instrument does not inform the user what system of measurement is in use! Induction disc type integrating meters are susceptible to error when connected to a severely distorted supply.

Power Factor Correction Capacitors

Any voltage distortion appearing across a power factor correction capacitor will cause harmonic currents to flow. The currents will be proportional to both the harmonic voltages and the harmonic numbers; they can be large and can damage the capacitors, particularly if the supply is being taken from a local generator.

Any lumped capacitance in the system may resonate at a harmonic frequency with transformer leakage reactance. Resonance causes large harmonic currents to flow in the components, and can lead to large harmonic voltages in parts of the system. Figure 4.5 illustrates the manner in which series and parallel resonances may arise.

The manner in which resonance is likely to arise is indicated by the following consideration of a 1-MVA three-phase 50-Hz distribution transformer with leakage reactance of 0.05 p.u. and a 240-V (phase to neutral) secondary winding:

$$\text{Leakage reactance} = \frac{\text{volts per phase} \times \text{p.u. leakage reactance}}{\text{rated current}} \quad (4.9)$$

$$= \frac{240 \times 0.05}{1389} = 8.64 \text{ m}\Omega \text{ per phase}$$

$$\text{Leakage inductance} = \frac{8.64 \times 10^{-3}}{2 \times \Pi \times 50} = 27.5 \text{ }\mu\text{H per phase}$$

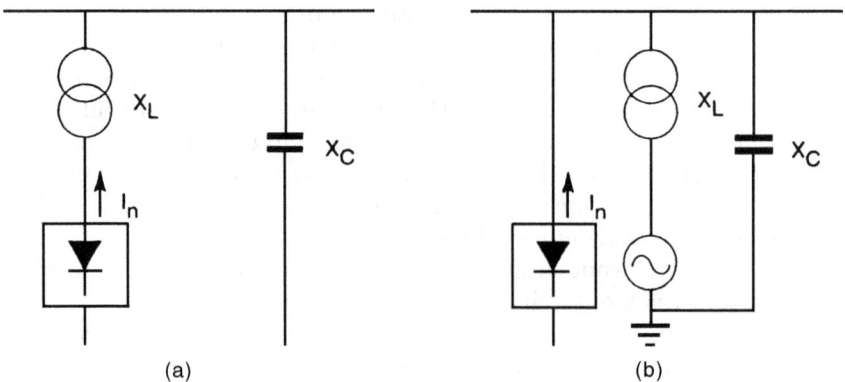

Figure 4.5 Examples of series and parallel arrangements of transformer leakage reactance and a capacitive reactance. (*a*) X_L and X_C in series. (*b*) X_L and X_C in parallel. The arrow I_N represents the harmonic current generated by the rectifier.

For resonance at the seventh harmonic:

$$\text{Capacitance} = \frac{1}{(2\,\Pi\,f)^2 \times L} \qquad (4.10)$$

$$= \frac{10^6}{(2 \times \Pi \times 350)^2 \times 27.5}$$

$$= 7519 \ \mu\text{F per phase}$$

This value of capacitor at 240 V and 50 Hz will result in a fundamental capacitor current of 567 A, and when operating as a three-phase bank the kVAr rating will be 408 kVAr.

A 400-kVAr capacitor bank would be used, for example, to correct the power factor of a 1-MVA load from 0.7 to 0.95 and is therefore likely to be encountered in practice.

Similar calculations undertaken for the eleventh and thirteenth harmonics lead to resonant capacitor ratings of 165 and 118 kVAr respectively. Such lower ratings may well be encountered if a capacitor bank is automatically switched in sections. A calculation for the fifth harmonic leads to a resonant capacitor rating of 800 kVAr, which would be unlikely to occur in practice.

Capacitors can be detuned by adding sufficient series inductance to ensure that the series circuit is inductive at the lowest significant harmonic (or the troublesome harmonic). The circuit will then be inductive at all higher harmonic frequencies.

Engineering Recommendation G.5/4

With these matters in mind the Electricity Association (of the United Kingdom) in 1976 issued Engineering Recommendation G.5/3, "Limits for harmonics in the United Kingdom Electricity Supply System." In 2001 a revised version was issued, Engineering Recommendation G.5/4, "Planning levels for Harmonic Voltage Distortion and the Connection of Non-Linear Equipment to Transmission Systems and Distribution Networks in the United Kingdom." G.5/4 is invoked by the United Kingdom network operators when a consumer wishes to install nonlinear equipment.

Document G.5/4 applies to all transmission and distribution systems from 400 V to 400 kV, different limits being applied to the various voltage systems. It applies to nonlinear loads which take power from the system, such as uninterruptible power supplies, and to generators arranged for parallel running which deliver power to the system; in the text no distinction is made between loads and generators, they are all regarded as nonlinear loads. The assessment procedure is in three stages:

Stage 1 allows the connection of equipment to 400-V systems without individual assessment provided that certain criteria are met. It applies to small equipment not exceeding 16 A per phase complying with BS EN 61000-3-2, and six-pulse and 12-pulse rectifiers not exceeding 12 and 50 kVA, respectively. Harmonic current emissions (up to the order 50) from aggregate loads are required to be within specified limits.

Stage 2 is applicable to installations supplied at a high voltage below 33 kV. It also applies to 400-V systems where a Stage 1 assessment is not appropriate due to the rating of the equipment, the emission levels, or the network characteristics. Harmonic current emissions from aggregate loads are required to be within specified limits. The voltage distortion caused by the new load is assessed and the effect of adding this to the existing distortion is predicted. The predicted distortion is required to be within specified limits.

Stage 3 is applicable to installations supplied at 33 kV and above. It also applies to installations at lower voltages which are found to be not acceptable under Stage 2. The procedure is somewhat similar to Stage 2, but the calculations are more rigorous and there are no harmonic current emission limits.

For Stages 2 and 3 the consumer has to provide the network operator with sufficient data to enable the appropriate calculations to be undertaken. If the consumer wishes to undertake calculations associated with Stages 2 or 3 the network operator will have to provide information relating to the supply system.

There is an associated guidance document, Engineering Technical Report 122 (ETR 122), which provides background information and includes additional information and worked examples.

The Point of Common Coupling

The point of common coupling is the point on the public electricity distribution system, electrically nearest to the consumer's installation, where other consumers are or may be connected. It is the point at which the network operator applies the limits defined in Engineering Recommendation G.5/4 and it is important that consumers understand the concept and the results of applying it.

The electrical distribution to premises usually falls into one of three patterns:

- The consumer is supplied at low voltage by a short length of feeder cable connected to a distributor cable which supplies other consumers. The point of common coupling will be regarded as the point of supply (the consumer's terminals).

- The consumer is supplied at low voltage by a dedicated cable from a nearby substation. The point of common coupling will be regarded as the substation low-voltage busbars.

- The consumer is supplied at high voltage and owns high-voltage switchgear, one or more transformers, and high- and low-voltage distribution systems as appropriate. The point of common coupling will be regarded as the high-voltage intake terminals.

The third pattern of distribution in which power is taken at high voltage is worthy of comment. The high-voltage network is likely to have a low impedance and the network operator will allow a large harmonic load to be connected to it, much larger than would be allowed if the point of common coupling was on the low-voltage side. If a large distorting load is connected to the low-voltage system the entire consumer's low-voltage system will be heavily distorted and damage and malfunction may result. As the point of common coupling is on the high-voltage side the network operator is neither responsible for nor concerned with distortion on the low-voltage side; any corrective action required will be the responsibility of the consumer.

If nonlinear equipment is to be installed within such a system the problem can be avoided by stating, within the contract documents, that Engineering Recommendation G.5/4 applies and the point of common coupling is to be regarded as the equipment terminals or some other clearly defined point within the low-voltage system.

The Effect of Bridge Rectifier Loads on Local Generators

Impedance of the Supply

If the electricity supply is derived from a local generator, all the effects mentioned in the previous section may be experienced but, as the source impedance will almost certainly be higher than that of a supply system, the effects are more likely to be troublesome. There are other additional effects which occur owing to the nature of synchronous machines, and these are described in subsequent paragraphs.

The magnitude of the effects described in this section depend on the rating of the local generator. If a rectifier is supplied from a very large generator the effects will be negligible and the supply may be regarded as a low-impedance system as considered in the previous section. However, many standby generating sets supply uninterruptible power supplies which have rectifiers as their input modules and the effects may well become relevant.

The Effect of Harmonic Currents

The simplified Eq. 4.8 for estimating harmonic distortion of a supply system indicates that the supply impedance determines the degree of distortion. While this remains true for a supply derived from a generator, a generator is less amenable to analysis and only a brief description will be attempted here.

An equivalent circuit for one phase of a synchronous generator is shown in Fig. 4.6. The rotor iron and the damper windings D_D and D_Q prevent rapid flux changes in the rotor, and any distortion of the voltage caused by harmonic currents will be mainly due to the impedance presented by the subtransient reactance X''. This may be regarded as the stator leakage reactance, approximately equal to the mean of the direct and quadrature axis subtransient reactances.

A method is therefore available to estimate the approximate voltage distortion due to each harmonic current. For a current I_n of harmonic order n the voltage across X'' will be :

$$V_n = I_n \times n \times X'' \text{ p.u.} \tag{4.11}$$

As an example, a fifth harmonic current of 0.1 p.u. passed by a machine having a subtransient reactance of 0.15 p.u. will produce a fifth harmonic voltage of 0.075 p.u. (7.5 percent of rated voltage). The total harmonic distortion can be estimated by performing this calculation for each significant harmonic current. The distortion can be reduced by reducing the machine subtransient reactance; this can be achieved by using a larger generator frame size or, if a purpose-designed machine is envisaged, by using more flux and fewer turns or a longer frame.

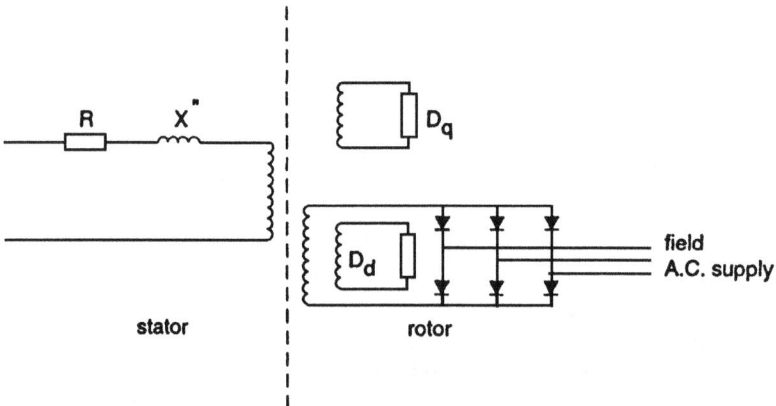

Figure 4.6 Equivalent circuit of a generator.

The Effect of Rectifier Commutation

It has already been stated that the period of commutation of a rectifier depends on the commutating reactance of its supply. If there is no rectifier transformer or series inductor the commutation period is determined solely by the supply system reactance. A normal supply system will have a low reactance leading to a short commutation time, and a small standby generator will have a comparatively high reactance leading to a longer commutation time. The longer commutation time leads to a reduction of dc output voltage and more pronounced notching of the supply voltage waveform.

When a three-phase generator supplies a steady linear load the stator currents produce a steady magnetic flux rotating, with the rotor, at synchronous speed. When the linear load is exchanged for a rectifier an entirely different set of conditions arises. The current will not be the idealized square waves previously considered (Fig. 4.2) but is likely to approach a trapezoidal shape due to the extended commutation periods. Between the commutation periods the stator flux remains stationary and moves forward in discrete steps as each commutation transfers current from one phase to the next.

With a diode rectifier this results in the type of distortion illustrated in Fig. 4.7. The magnitude of the distortion will depend on the reactance of the supply, as already discussed.

With a thyristor rectifier the operation is similar to that of diodes but the depth of the notching will be deeper depending on the commutation delay angle and, in idealized form is illustrated in Fig. 4.8.

When running from a local generator, owing to the irregular rotation of the stator flux the notching in the succeeding phase can be considerably deeper (and the rise in the preceding phase correspondingly less)

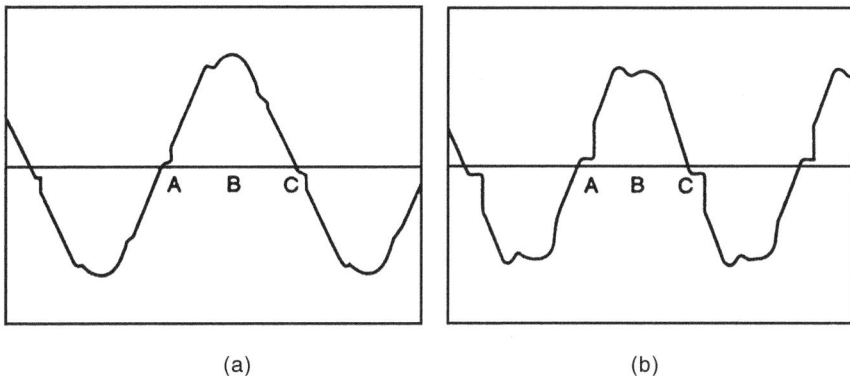

(a)   (b)

Figure 4.7   Typical distortion caused by rectifier loads on a local generator. (*a*) 50 percent load. (*b*) 100 percent load.

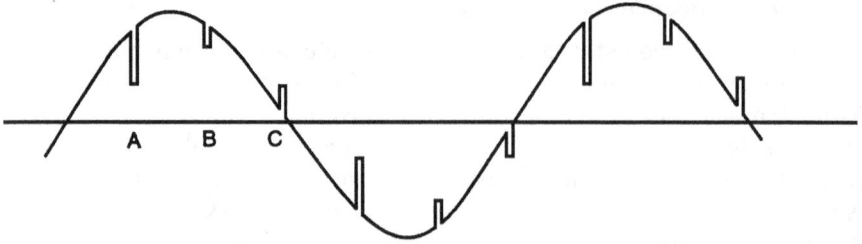

Figure 4.8 Typical distortion caused by thyristor loads on a local generator.

than that experienced from a low impedance supply. Extreme cases of this type of distortion are illustrated in Fig. 4.9.

An interpretation of Figs. 4.8 and 4.9 may be useful. At time *A* the voltage experiences a negative notching while current is being transferred from the preceding phase. Conduction continues for 120°, and ends at time *C* when the voltage experiences positive notching while current is being transferred to the succeeding phase. In Fig. 4.9 the notching is severe and may be exaggerated by ringing of a measuring or other local circuit.

Additional notching appears at *B*, the midpoint of the conduction period between *A* and *C*; this occurs at the time when the other two phases are commutating. This is a phenomenon that arises when a bridge rectifier is supplied from a generator with transient saliency. During commutation the stator flux advances rapidly to its new position, and in doing so induces reverse voltages in the stator windings. Any generator driven by a diesel engine and supplying an uninterruptible power supply is likely to have a salient pole rotor and is susceptible to this effect. Theoretically, the effect should be much reduced by extending the damper windings over the interpolar regions and, in practice, the effect is lessened by interconnecting the damper windings across the interpolar gaps.

If a generator is shunt excited (see Fig. 1.5) any notching of the stator voltage appears in the excitation circuit and can cause additional distortion. The use of a separate shunt excited ac exciter reduces the possibility, but the best arrangement is to use a permanent magnet pilot exciter followed by a main exciter (see Fig. 1.4).

Additional Rotor Losses

Harmonic stator currents drawn by a bridge rectifier cause air gap fluxes of the same general shape as the fundamental but rotating at *n* times synchronous speed, where n is the harmonic number. These will induce currents in the rotor iron and windings, adding to the rotor losses and increasing its temperature.

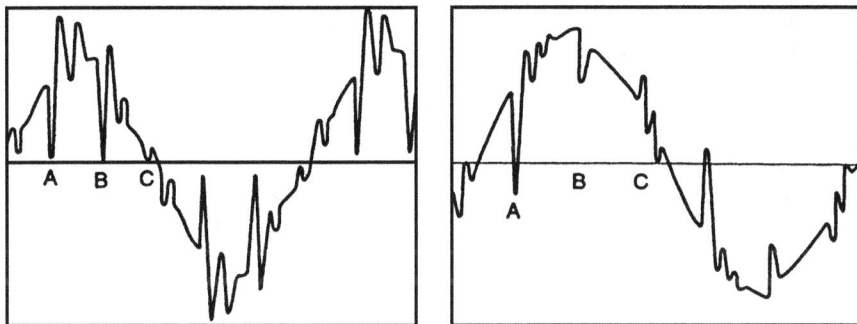

Figure 4.9 Examples of extreme distortion caused by thyristor loads on a local generator.

Rectifier harmonic currents occur in pairs such as the fifth, with negative phase sequence, and the seventh, with positive phase sequence. Relative to the rotor the fifth will rotate in the opposite direction at six times synchronous speed, and the seventh will rotate in the same direction as the rotor, also at six times synchronous speed. These two flux systems with opposite rotation result in an alternating flux that is stationary with respect to the rotor. The alternating flux induces damper winding currents that are stationary relative to the rotor, causing localized extra heating. Within the rotor the effect is similar to that caused by single-phase or unbalanced loads. Most machines are fairly tolerant of unbalanced loads and overheating is unlikely unless the rectifier loading (six-pulse) exceeds say 50 percent of the generator rating. Machines with laminated pole faces are recommended for such duties.

Torque Pulsations

The alternating flux mentioned in the preceding paragraph and the fundamental stator flux are both stationary relative to the rotor. The reaction between these two fluxes results in a pulsating torque, the magnitude being dependent on the spatial relationship between them. The spatial relationship is dependent on the machine parameters and the commutation delay angle.

The largest torque pulsations for a six pulse rectifier are at six times the supply frequency, and for a 12-pulse rectifier at 12 times the supply frequency. They rarely attain a significant magnitude and are not often troublesome in practice. The coupling between the flywheel and the alternator will include some damping mechanism, and the effect seen by the crankshaft will usually be small compared with the torque irregularities it experiences in service. Where the rectifier loading is a significant part of the rating of the set it would be wise to ensure that the lowest natural frequency of the rotary system is well below the lowest forcing frequency (which will be equal to the rectifier pulse number).

Effect on Electronic Devices

A local generator is likely to incorporate electronic devices such as a tachometer and a voltage regulator which may use the zero voltage crossing points as timing signals. Deep notching of the supply voltage can lead to additional zero crossings and can affect the operation of such devices. If problems are experienced it may be possible to obtain a clean supply by using a small low-pass filter.

Voltage regulators may be affected by a distorted supply; if the regulator is expected to control the fundamental component of voltage, the harmonic voltages must be removed from the sensing signal and a low-pass filter may achieve this as suggested in the previous paragraph. If the harmonic voltages are not removed the regulator will set to the true rms or the average value depending on the design. In cases of severe distortion this can lead to considerable errors.

Reduction of Distortion Due to Rectifier Loads

Increasing the Pulse Number

In the section titled "Harmonics Generated by Bridge Rectifiers" it is stated that the current taken by a three-phase six-pulse bridge rectifier includes only the fifth and seventh, eleventh and thirteenth, seventeenth and nineteenth, etc., harmonics. If the rectifier pulse number is increased to 12, the fifth and seventh, seventeenth and nineteenth, etc., are canceled while the magnitudes of the remaining harmonics are unaltered. To comply with the planning levels of Engineering Recommendation G.5/4 it is often necessary to use 12-pulse rectifiers.

For the same reason the supplies to rectifiers of multiunit UPS installations should be phase shifted, to achieve a higher pulse number. The supplies to two paralleled units should be shifted by 30°, three units by 20°, etc. For a phase-shifted multiunit redundant system, the loss of a unit will result in some additional harmonic currents being drawn from the system.

The manner in which this harmonic cancellation occurs is not obvious and the reader may wish to study Table 4.3 which is an attempt to explain how it is achieved, it is not offered as a rigorous mathematical exercise.

It can be seen that for both fifth and seventh harmonics the two secondary currents have a phase difference of 180°. It follows that they will be cancelled in the primary; similar reasoning may be applied to the seventeenth and nineteenth harmonics.

TABLE 4.3   The Manner in Which Harmonic Current Cancellation Is Achieved in a 12-Pulse Rectifier

Phase shifting transformer arrangement		
Fundamental secondary voltage shift relative to primary voltage	-30^0	0
5th harmonic current shift relative to primary voltage	$5 \times (-30^0)$ $= -150^0$	0
5th harmonic current has -ve sequence: Shift due to transfer from secondary to primary	-30^0	0
5th harmonic current total shift from rectifier to primary	$-150^0 -30^0$ $= -180^0$	0
7th harmonic current shift relative to primary voltage	$7 \times (-30^0)$ $= -210^0$	0
7th harmonic current has +ve sequence: Shift due to transfer from secondary to primary	$+30^0$	0
7th harmonic current total shift from rectifier to primary	$-210^0 +30^0$ $= -180^0$	0

Adding an Input Filter

It is possible to reduce the harmonic currents flowing through the supply system by providing a low-impedance path across the rectifier power input terminals. The harmonic currents generated by the rectifier will then be shared between the supply system and the shunt path in a manner dependent on the inverse ratio of their impedances.

The low-impedance path may comprise one or more shunt connected filter circuits tuned to particular harmonics, or a simple untuned capacitor. In either case a series inductor may be included on the supply side, which effectively increases the system impedance, renders the shunt path more effective, and reduces the harmonic current flow through the supply system.

The series inductor may also serve another purpose; if properly chosen it can prevent any likelihood of the capacitor (or the net capacitance of the filter) resonating at a harmonic frequency with, for instance, the leakage reactance of a local transformer. This subject is discussed in connection with power factor correction capacitors in the section titled "The Effect of a Bridge Rectifier on a Supply System."

Shunt-connected harmonic filters and capacitors will take a leading fundamental frequency current which may result in the UPS operating at a leading power factor, particularly at light loads, and almost certainly if a diode rectifier is involved. At the expense of some complication it may be possible to install a capacitor bank with automatic switching, but there may well be conflict between the requirements to maintain harmonic attenuation and to control power factor.

Active Power Filters

This is a system of "cleaning up" a distorted current wave by continuously monitoring the wave shape of the current taken by the load, comparing it to a sine wave, and injecting into the upstream system a current equal to the distortion. It follows that the filter equipment supplies the distorting harmonic currents which do not have to be taken from the supply. Active power filters became widely available in the 1990s and may well become more widespread in the next decade. They use inverters to generate the complex wave shapes, for a particular installation the rating of the inverter will be related to the degree of correction required. It follows that there will be additional energy consumption, related to the degree of correction, the cost of which must be taken into consideration.

Pulse Width Modulation of the Input Current

Instead of taking the current in blocks at the rate of one block per half cycle the current is taken in short pulses occurring at a high frequency which may be up to tens of thousands of hertz. The width of the pulses is continuously controlled so that they are widest at the center of each half cycle and narrow towards the zero crossover points. The high frequency components are removed by a shunt-connected filter and only the fundamental component of current is drawn from the supply. The power factor (cos φ) is determined by the pattern of pulses and the equipment is normally arranged to operate at unity power factor. UPS equipment incorporating input rectifiers of this type became available in the late 1990s.

Switched Mode Power Supplies

This is a rectifier configuration which is universally used in electronic equipment and is infamous for the peaky current which it takes, it is said of the current that the fundamental cannot be seen for the harmonics that surround it. The basic power supply comprises a series arrangement of a single-phase rectifier, a shunt-connected capacitor, a chopper with associated control circuit, and an output filter. The rectifier and capacitor produce a peaky current flow in the middle of each voltage half cycle—a waveform rich in harmonics particularly the third. A typical current waveform is illustrated in Fig. 4.10.

The waveform of Fig. 4.11 is a useful approximation of Fig. 4.10, it is made up of the following harmonic components:

Harmonic order (n)	1	3	5	7	9	11	13
Magnitude (%)	100	−85	60	−30	18	−12	8

Triplen harmonic currents in a three-phase system have zero phase sequence and become additive in the neutral (if there is no neutral conductor there can be no triplen harmonic currents). In an installation that includes a large proportion of IT equipment there are likely to be large triplen harmonic currents and these must be taken into consideration at the planning stage. These currents may flow in the UPS neutral, the UPS bypass neutral conductor, the neutral conductors in the distribution system, and the standby generator neutral.

Triplen harmonic currents may be removed from a distribution system by providing a low-impedance shunt path in the form of a transformer with a delta winding which can be either the primary or a

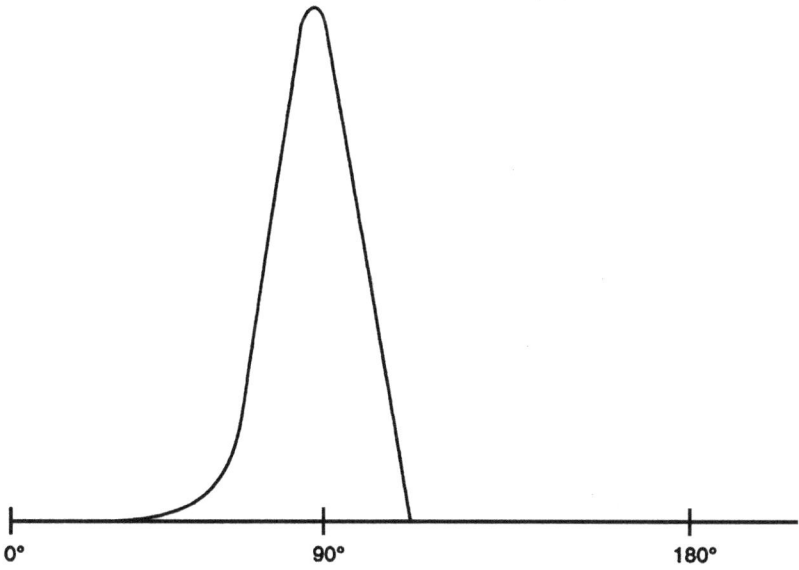

Figure 4.10 Typical waveform of current taken by a switched mode power supply.

a_1 =	1.00
a_3 =	- 0.85
a_5 =	0.60
a_7 =	- 0.30
a_9 =	0.18
a_{11} =	- 0.12
a_{13} =	0.08

```
0   20        30      50      70      90     110     130     150     170
```

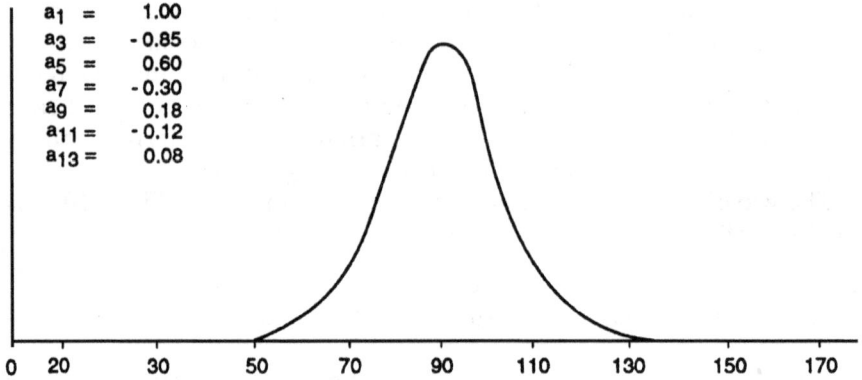

Figure 4.11 Approximation to the waveform of current taken by a switched mode power supply.

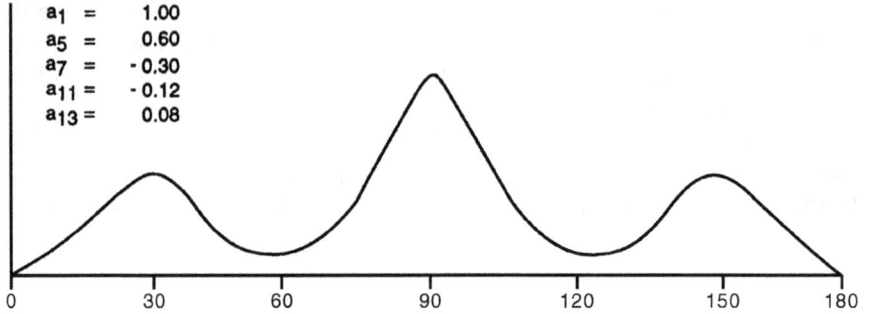

a_1 =	1.00
a_5 =	0.60
a_7 =	- 0.30
a_{11} =	- 0.12
a_{13} =	0.08

```
0            30         60         90         120        150        180
```

Figure 4.12 The waveform of Fig. 4.11 without the triplen components.

tertiary winding. The load-generated triplen harmonic currents will flow in the transformer secondary. Core fluxes will cause triplen currents to circulate in the delta winding, and these triplen currents will neutralize those in the secondary; significant triplen voltages will not therefore appear at the primary terminals.

If the triplen harmonic currents are removed from the waveform of Fig. 4.11, the shape is changed to that shown in Fig. 4.12 which includes the following harmonic components:

Harmonic order (n)	1	5	7	11	13
Magnitude (%)	100	60	-30	-12	8

The new waveform is far from ideal but it is a considerable improvement; the peak current has been reduced by one-third, the period of conduction has been extended to the complete half cycle, and the rms value has been reduced by 19 percent.

Bibliography

British and European Standards

BS EN IEC 61000—Electromagnetic compatibility
Part 3-2 Limits for harmonic current emissions (equipment input current up to 16 A
per phase)

Other Documents

Engineering recommendation G.5/4—Planning levels for harmonic voltage distortion
and the connection of non-linear equipment to transmission systems and distribution
networks in the United Kingdom, The Electricity Association of London, United
Kingdom.
Engineering technical report 122—Guide to the application of Engineering
Recommendation G.5/4 in the assessment of harmonic voltage distortion and connec-
tion of non-linear loads to the supply systems in the United Kingdom, The Electricity
Association of London, United Kingdom.
Yacamini, R., "Power system harmonics," A series of four tutorial articles in *IEE Power
Engineering Journal*, The Institution of Electrical Engineers of London, United
Kingdom, August 1994, February 1995, October 1995, and August 1996.

Static Uninterruptible Power Supplies

Definition

A static UPS system is a circuit which ensures a continuous power supply to the load irrespective of outages , spikes, brownouts, or other disturbances from the normal incoming mains supply. It is achieved by using solid-state circuitry which employs a battery or possibly kinetic energy as the alternative energy source.

Background

The development of static UPS clearly was dependent on the availability of solid-state switching devices. The earliest conversion systems available in 1960 for dc to ac were no more than mechanical vibrators with ratings no higher than 500 VA used for radio/communications applications. The advent of power transistors enabled the first true static inverters to be built, applications in the early years being communication and instrumenation. Thyristors then became available and, gradually, ratings of modules increased. It should be remembered in these early days that switching devices suffered from wandering characteristics due to operating temperature and aging. In time these problems were solved. By approximately 1960 computers began to require UPS systems

From inception to todays designs we have seen dramatic improvements, efficiency originally at 80 to 81 percent compares with claimed efficiencies of up to 98 percent for present designs. The size of systems also has reduced considerably, modern designs are now some 60 percent less in proportion. Maintenance costs are much reduced and reliability figures now quoted are some 10 years mean time between failure for a single module. Note this figure does not allow for battery reliability. Figures for a multimodule parallel redundant system are subject to variations and circuit complexity, but as a guide 20 years is achievable.

Basic Design

A typical basic circuit is shown in Fig. 5.1. Under normal operation power to the load is fed through the circuit, in other words, rectifier/battery charger, inverter load. On unavailability of the mains supply for any reason then the load is fed from the battery/inverter.

The static switch is present for two important reasons:

1. *In the event that surge loads appear, for example, starting currents from some devices which are part of the load, or indeed the rupturing of subcircuit fuses which will demand high currents:* Under these conditions it is highly probable that the inverter output will be unable to meet the power demand and voltage will rapidly decrease. Such a condition will ensure that an alternative supply is made immediately available via the static switch from the normal mains supply. The static switch reverts to inverter supply for the load when the surge subsides.

2. *The static switch also ensures power to the load if a failure in the system occurs.*

This module as described is typical, and has been in use for many years, it employs basic subassemblies, namely, rectifier/charger, inverter, static

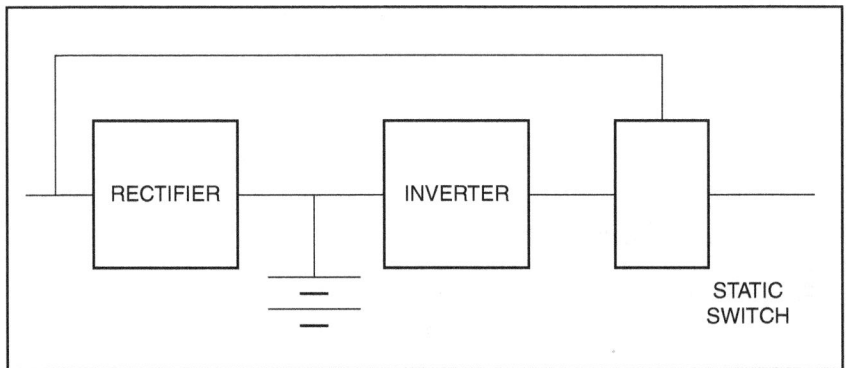

On Line

Figure 5.1

switch, which may be used as subassemblies in many other circuits each having its own properties and uses. These will be described in more detail later in this chapter.

Subassemblies

Before detailing the various alternative circuits, a general description of the subassemblies is given below, commencing with the rectifier/battery charger.

This subassembly is designed to provide full-load power through the inverter assembly , recharge a battery, and also cater for losses in the system.

Rectifiers

As an example, a fully controlled six-pulse thyristor bridge is normally used as shown in Fig. 5.2. Note the conduction point is varied to control the dc output voltage. L in the diagram represents an impedance to attenuate ripple which will affect battery performance.

The rectifier bridge offers the ability to ramp up power on start-up, thus affording excellent loading conditions to a supporting diesel generator set. Ramp up may be set from No Load to Full Load within 5 to 30 s.

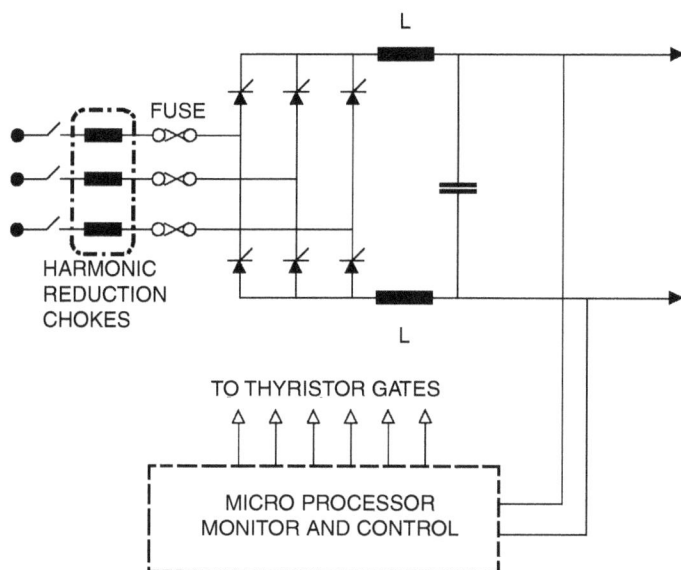

Figure 5.2

Battery characteristics impose certain design conditions on the rectifier charger, assuming that a *valve regulated lead acid* (VRLA) battery is used then the normal float charge level is 2.27 volt per cell (VPC), and at the end of discharge the voltage level may be as low as 1.6 VPC. The end of discharge voltage is dependent on the autonomy period and design of the battery, guidance from the battery manufacturer is required. There is the possibility that the system may require to be fitted with a boost charge facility, this may be required as an initial charge when commissioning the module, this voltage will be 2.3 to 2.4 VPC. Such a requirement will be switched on manually and also switched off either manually or shut down by a timer. Boost charging may be required for flooded-type lead acid and nickel-cadmium cells.

The rectifier/charger will also need to ensure that the ripple on dc output is within specified limits. Battery manufacturers limit the ripple to 7 percent of the C3 (3-h) capacity or 5 percent of the C10 rating under these conditions temperature rise within the cell will not exceed 5°C. In practice a 1-percent voltage ripple is normally accepted. Ripple does affect battery life, optimum ripple damage occurs between 100 to 250 Hz which normally with modern design UPS modules does not cause a problem.

The rate of charge is also important and in practice a current limit is set, thus the dc voltage gradually increases to the normal float charge level. The accepted standard for recharging a battery is to 95 percent capacity within 10 times the discharge period. If this requires shortening, the solution may be to increase the *ampere hour* rating of the cell thus avoiding exceeding the current charging limitation.

Other types of battery may require variations to the above characteristics. Details are given in Chap. 7.

Harmonics and Effect on Design

Clearly rectifiers will induce harmonic distortion onto the incoming supply. Such harmonics are the cause of concern to power networks and limitations are imposed to attenuate such problems. Electricity Council G5/4 gives recommendations designed to limit the harmonics in a distribution network to below those thought to cause EMC problems. In effect it recommends 5 percent total harmonic distortion (THD) as a maximum. There are some exceptions to the rule: For example, excursions outside the limit for limited time periods may be allowed, and the recommendation is not retrspective. It should be noted that this recommendation supersedes an earlier paper G5/3 which allowed a 10 percent THD for ratings above approximately 100 kVA. In the United States the standard now is IEEE 519 (1992) which aims for 5 percent THD. In Europe a similar standard is being considered EN 6100-3/2 which is in a trial period at the moment.

Figure 5.3 illustrates the harmonics to be expected with 6- and 12-pulse modules and the effect of running such units in parallel with some phase-shifting transformers.

Figure 5.4 shows wavetraces using various six-pulse modules with phase-shifting transformers, and an alternative fifth harmonic filter.

A fifth harmonic filter has financial advantages compared to a 12-pulse rectifier but has some drawbacks. The filter is a tuned circuit, and variations in frequency which may occur when operating with a diesel generator set may be a problem. Standard diesel set specifications allow the frequency to be controlled to ± 8 percent in which case it may result in detuning of the filter, producing higher harmonics. However, most installations use a close tolerance diesel generator set giving ± 1 or 2 percent frequency regulation thus obviating the problem. Another cause for concern is the dynamic response of modern diesel sets to a sudden load change, which may also affect output frequency briefly. Low load conditions may also affect the efficiency of the filter

The alternative is to use an active filter network, in essence a small inverter circuit using a high switching speed (20 kHz) producing *pulse width modulated* waveform. This is injected into the rectifier waveform

Number of Rectifiers	6 Pulses	1	2	3	4
	12 Pulses	–	1	–	2
	H3	1%	1%		1%
	H5	41%	3%		3%
	H7	11%	1%		1%
	H11	7%	7%		1%
CURRENT	H13	2.5%	2.5%		1%
HARMONIC	H17	2.5%	–		–
DISTORTION	H19	2%	–		–
	H23	2.5%	2.5%		2.5%
	H25	1%	1%		1%
	H29	1%	–		–
	H35	1%	1%		–
	H37	1%	1%		–
	THD*	43%	9.5%		5%

Figure 5.3

	without filter		with rectifier choke or transformer		with 5th filter	
	Half Load	Full Load	Half Load	Full Load	Half Load	Full Load
H3		1%		1%		1%
H5		41%		28%		3%
H7		11%		2%		4%
H11		7%		7%		7%
H13		2.5%		2.5%		2.5%
H17		2.5%		2.5%		2.5%
H19		2%		2%		2%
H23		2.5%		2.5%		2.5%
H25		1%		1%		1%
H29		1%		1%		1%
H35		1%		1%		1%
H37		1%		1%		1%
THD*	59%	43%	44%	30%	18%	11%

*THD = Total Harmonic Distortion

Figure 5.4

to produce a virtual sine wave. This is illustrated in Figs. 5.5 and 5.6. Such units are available as stand-alone modules; they do add approximately 1 to 2 percent to overall system losses.

Inverters

Inverter systems have advanced over the years and are the major area contributing to advances in system reliability and efficiency. The basic design is a simple bridge switching circuit (see Fig. 5.7a). Clearly the switches utilized a variety of solid-state switches and as such devices increased in performance so the inverter developed. In Fig. 5.7b the earlier circuits resulted in a square wave output clearly requiring a large filter to obtain a sine wave. This entailed losses in the filter and a poor dynamic performance. With developments in circuitry and availability of switching devices step wave systems were developed that have an increase in efficiency due to a smaller filter and an increase in system dynamic performance.

Early designs used power transistors and thyristors, and nowadays a device known as an *insulated gate bipolar transistor* (IGBT) is in com-

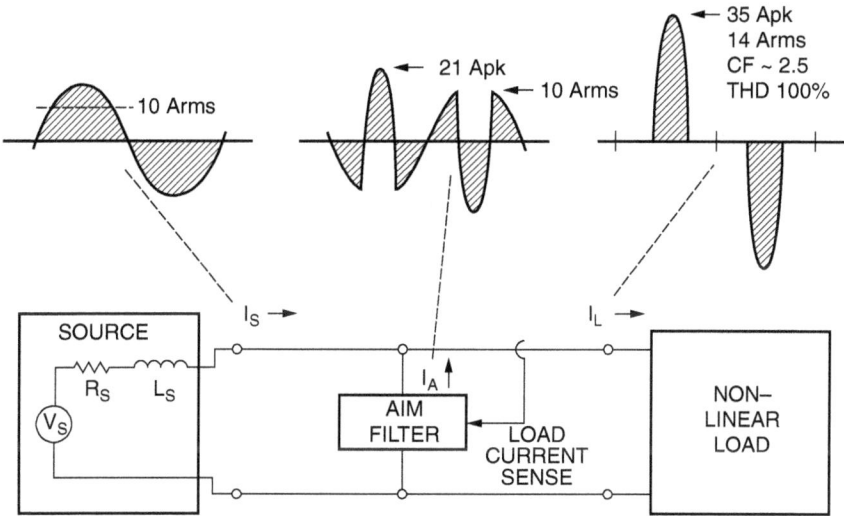

35 Apk
14 Arms
CF ~ 2.5
THD 100%

21 Apk
10 Arms

10 Arms

$I_S \rightarrow$

$I_L \rightarrow$

SOURCE

R_S L_S

V_S

$I_A \uparrow$

AIM
FILTER

LOAD
CURRENT
SENSE

NON–
LINEAR
LOAD

Figure 5.5

The three waveforms below show the performance of the AIM Filter.
The left hand waveform is a "harmonic-rich" load current waveform that is typical
of "problem" non-linear loads.
In the center is the current waveform from the AIM Filter, which when combined with the
harmonic load current, produces a substantial cancellation of the offending harmonics
resulting in the desired smooth sinusoidal current waveform from the distribution system.
In this particular case, the harmonic content has been reduced from approximately
78 percent to 6 percent, a reduction of 13 to 1.

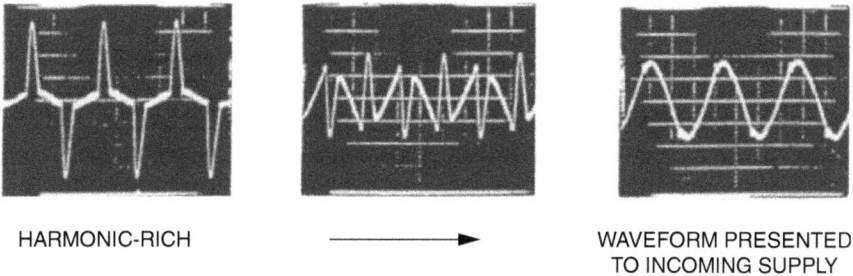

HARMONIC-RICH   ⟶   WAVEFORM PRESENTED
                                                              TO INCOMING SUPPLY

Figure 5.6

mon use. This device is the result of the combination of the properties
of the bipolar transistor and the MOSFET.

The bipolar transistor has advantages of high current and high voltage characteristics and the MOSFET of speed and single gate drive

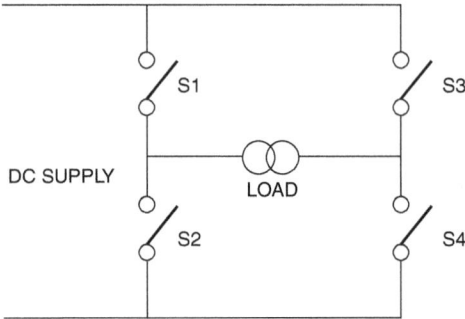

DC SUPPLY

(a)

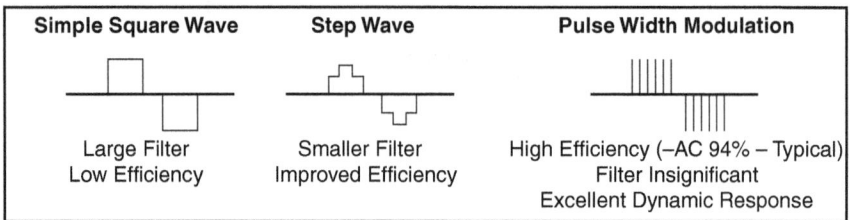

(b)

Figure 5.7

requirements. The resultant device IGBT now possesses high switching characteristics and good voltage control. Figure 5.8 illustrates a typical bridge circuit, and the output waveform. Note that the heavy black waveform indicates the output and the true sine wave the output from the very small filter.

The IGBT module provides a high-speed switching system using PWM waveform. Switching speeds vary between manufacturers usually between 3 to 30Khz. The choice of switching speeds is governed to a great extent by two side effects: The higher the switching speed the smaller the output filter required to obtain a good sine wave, and, conversely, high switching speed tends to evolve high EMC disturbances. Also high switching speeds tend to increase losses and heat output.

There is no doubt that IGBT designs have had a significant effect on the output filter size, thus reducing the overall size of the module. Dynamic performance has improved no load–to–full load and its converse, being in the order of of ±5 percent and returning to 1 percent within 40 ms. Additionally the circuit has improved the ability to cope with crest factors, a typical UPS now being able to support a crest factor 3:1 at full load.

The inverter is designed to cope with a variety of loads, computers and communication equipment being the predominant application.

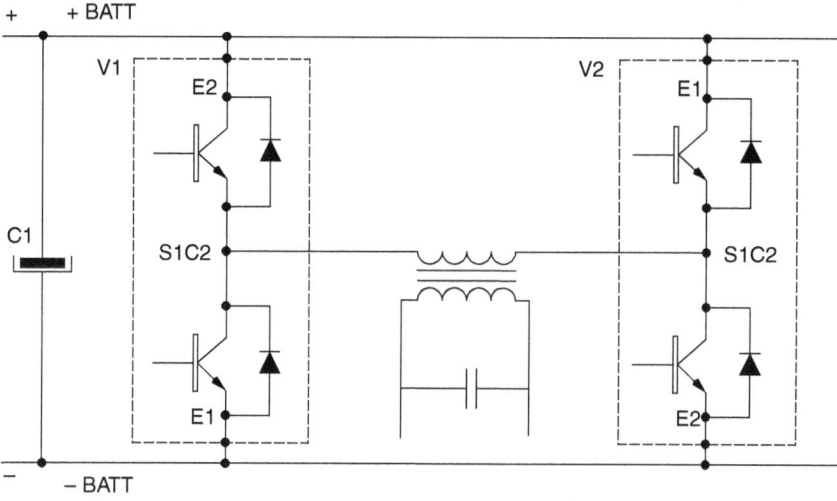

Working Principle of the IGBT Inverter Bridge

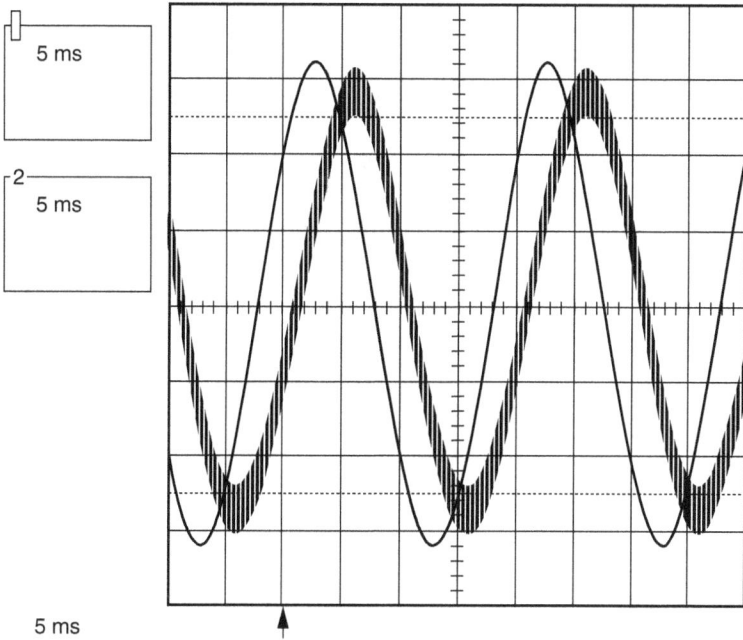

Output Voltage / Inverter Bridge Current at Np

Figure 5.8

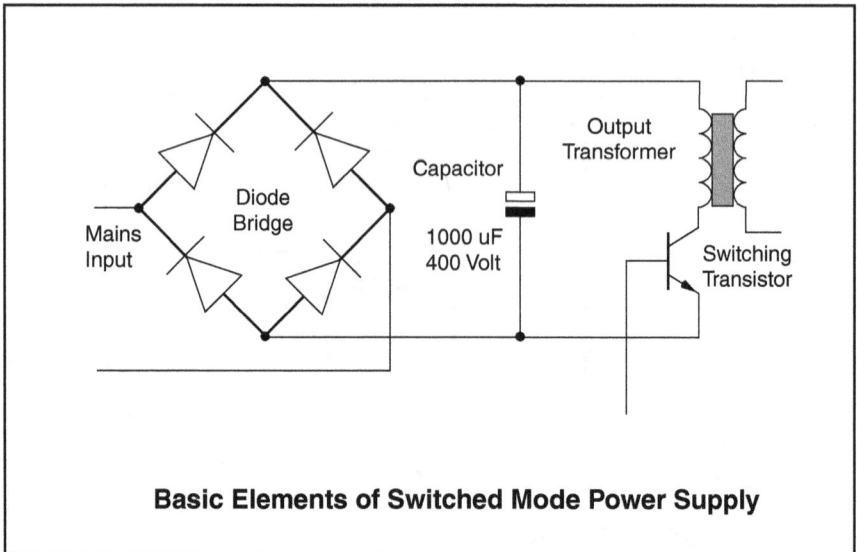

Basic Elements of Switched Mode Power Supply

Figure 5.9

Computers, although demanding ac, in fact operate internally on 3 to 5 volts dc, with very onerous voltage/ripple variations allowable. To comply with this requirement historically, large rectifier transformers with smoothing circuits were employed. Such devices rapidly approached the volume of early computer systems. Therefore, to reduce size and losses 400 Hz was a chosen input to the computer. This assisted in providing a supply which significantly reduced the size of the system and also gave good isolation from external supply variations.

Nowadays, the switch mode power supply has assumed a predominant role. This device is compact and has a relatively high efficiency (see Fig. 5.9). However, this device does have a peaky waveform (see Fig. 5.10). Measurements and wavetraces are available showing waveforms with 6:1 crest factor, but for most design considerations due to the induction of impedance and capacitance effects by cable and distribution effects a 3:1 crest factor is seldom exceeded.

Types of Loads

Computer loads in the past were frequency sensitive , modern design has obviated this problem to a large extent,and typically such loads will easily operate on ±3 to 5 percent frequency variation.

Consideration of other loads has to be examined. For example, lighting loads can induce their own problems. The incandescent lamp has a

CURRENT WAVEFORM OF A TYPICAL MINI-COMPUTER

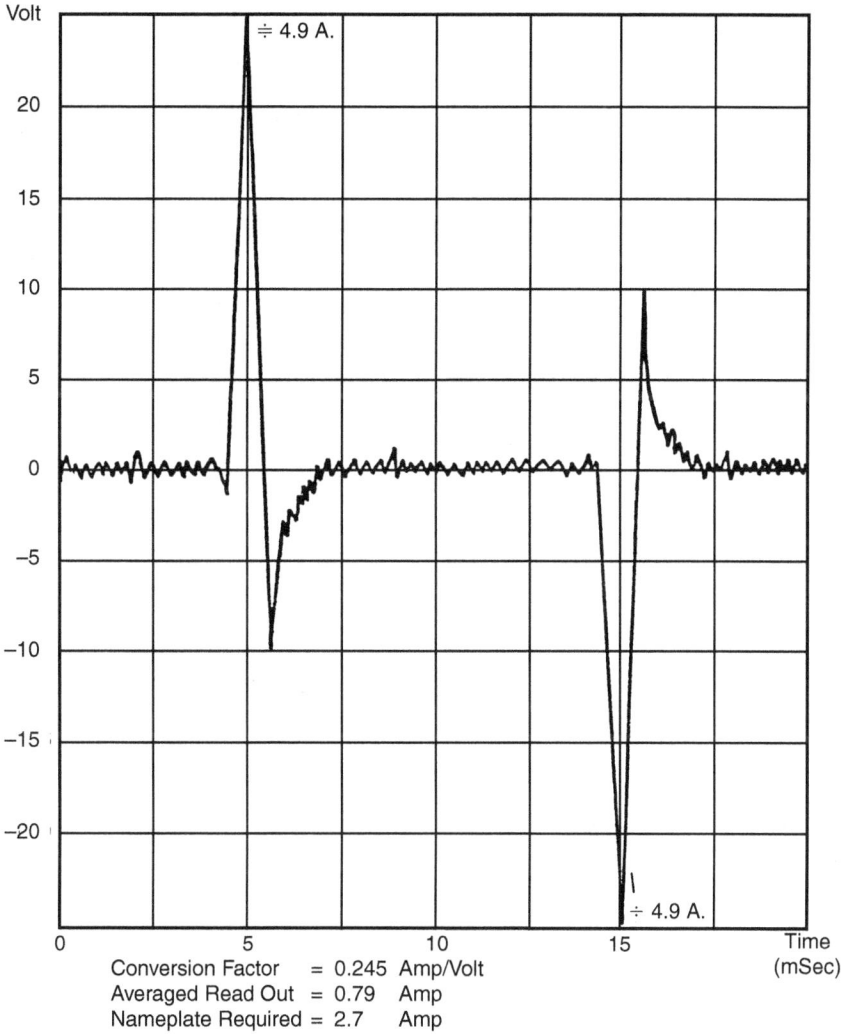

Conversion Factor   = 0.245 Amp/Volt
Averaged Read Out = 0.79   Amp
Nameplate Required = 2.7   Amp

$$\text{Crest Factor} = \frac{4.9}{0.79} = 6.2$$

Figure 5.10

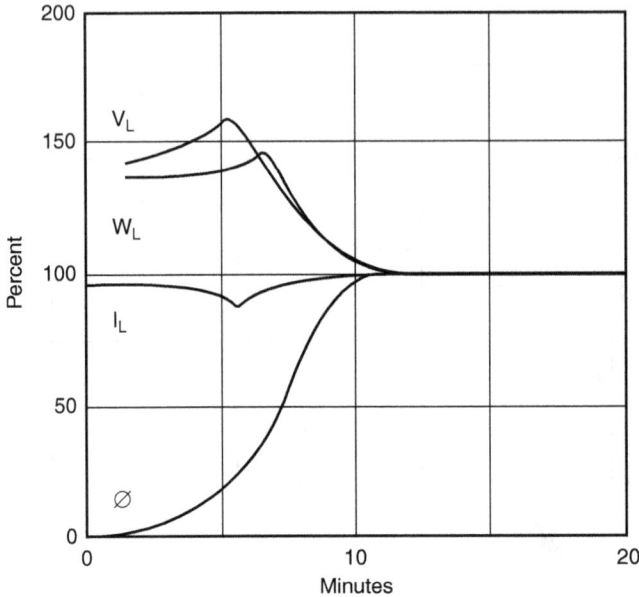

Typical Lamp Start Condition
Sodium Discharge Lamp.
V_L = Lamp Voltage
W_L = Lamp Wattage
I_L = Lamp Current
$\varnothing$ = Lamp Lumin Average

Figure 5.11

high starting surge, depending on the type of lamp between 15 to 20 times normal run current decaying to normal load current within 20 to 100 ms. Discharge lamps have unusual starting conditions (see Fig. 5.11). *O* represents light output, *IL* lamp current, *W1* lamp wattage, and *V1* lamp voltage. These figures are lamp figures, lamp control gear present the incoming supply with approximately two times full load current, decaying within 10 min for sodium lamps. These figures of starting time vary with lamp type.

Motor loads can present the power supply with a wide variety of starting surges, for example, squirrel cage motors giving 7 to 10 FLC with a time span for decay to FLC being anywhere between 1 to 10 s.

Typical Specification

As a guideline a modern UPS output will provide the following performance figures:

Voltage regulation ±1.5 percent

Voltage regulation with 100 percent load variation ±5 percent returning to nominal within 40 ms

Overload 110 percent for 1 h

Overload 120 percent for 10 min

Overload 150 percent for 1 min

Exceeding overload conditions necessitates the system reverting to mains operation via the static switch

Crest factor 3:1 at full load

Frequency 50 or 60 Hz when working with internal oscillator (independent of mains) ±0.01 percent

When working with mains system will accept ±3 percent outside these set tolerance figures will revert to working on internal oscillator which means the static switch is inoperable.

Frequency slew rate ½ Hz per s

For three phase systems:

Voltage phase shift with

Balanced load   120 ± 1 degree

Unbalanced load   120 ± 2 degrees

Efficiency at full load 93 percent; at 50 percent load 92 percent

Ambient temperature of operation 0 to 40°C.

Relative humidity (no condensation) 95 percent

Maximum altitude 1000 m

Noise level nowadays is much improved due to the use of modern switching techniques and wound components and you may expect 70 to 75dba at 1 m from the unit for systems up to 1 MVA.

The dc voltage for UPS systems depends for large systems on the incoming mains supply in most cases. For example, 400-V three-phase supply would use 180 cells, giving a float voltage of 410 V.

Static Switches

Static switches are used as shown in Fig. 5.1 to provide a virtually instantaneous transfer of power between supplies.

Most modules now use straight static switches with a mechanical switch in parallel (see Fig. 5.12).

Under normal operation the power is fed via rectifier/inverter static switch to the load, with the mains frequency controlling the operational

CONTACTOR
K50

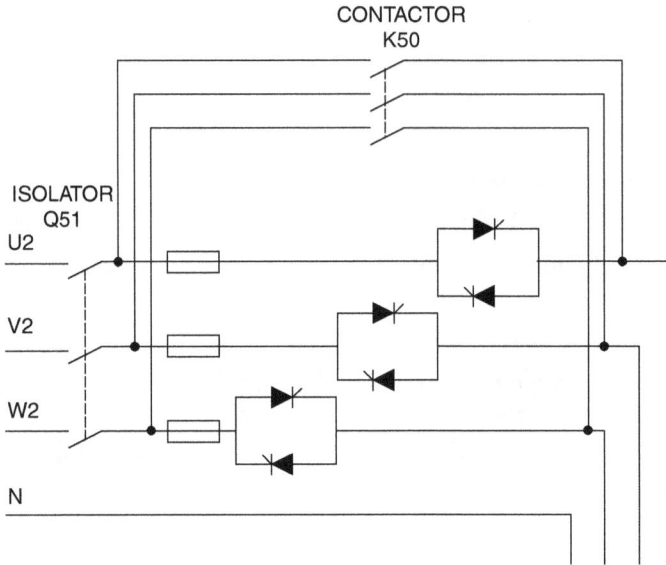

Figure 5.12

frequency of the inverter. Thus, the mains and inverter output are in synchronism.

If mains frequency is out of tolerance, in other words, outside ±3 percent, then the inverter frequency is controlled by its natural frequency, thus the static switch is inoperable.

The static switch transfers load from inverter to mains under the following conditions:

1. Surge loads causing an inverter overload will induce static switch operation, on departure of the surge the static switch will transfer load to inverter operation.

2. Failure of the inverter will immediately induce static switch operation.

3. Manual operation of the switch may occur, for example, during maintenance operations.

4. Operates at end of battery backup time

Various figures are quoted for the operational performance of the static switch. Transfer can be assumed to occur when voltage is out of tolerance by ±10 percent and transfer time is in the order of 500 μs.

The brief descriptions of the UPS subassemblies above are applied to a variety of circuits all of which are identified as UPS systems. Block diagrams and comments on some these circuits are given in the following section.

Designs Now Available

The full on-line system shown in Fig. 5.1, now also referred to as a *double conversion* module, is better illustrated in Fig. 5.13. Under normal operation, that is, mains supply is present, the oscillator firing the bridge circuit IGBTs will accept a signal from the mains waveform, ensuring that the system is in synchronism with that supply. Thus, the static switch can provide an alternative supply which is in synchronism with the supply to the load. If, for any reason, the mains frequency is unstable, then the oscillator control will break away and not accept mains frequency. Clearly, mains frequency acceptance level may be set on site and usually this is set at ±3 to 4 percent.

The system has the advantage of affording complete isolation from mains supply and the module will operate over quite large variations in input supply. The module is normally fitted with a maintenance by-pass, enabling maintenance without disturbing the load. Nowadays, the module is used on ratings of 10 kVA and above, and is comparatively expensive. Further developments have allowed dispensation of the input transformer reducing weight and losses. In addition, the rectifier may now employ IBGT switching techniques, enabling a very large reduction in the value of harmonics. Such systems now easily meet the latest guidelines on harmonics. It should be noted that the employment of IGBT switching systems in the rectifier does increase cost.

Another circuit described as delta conversion (Fig. 5.14a) consists of two inverter chargers in series, battery, and a transformer in line. Under normal operation and assuming that mains is at nominal voltage the load is fed directly via the primary winding of the transformer.

Inverter 1 is only supporting the mains current, which clearly in this case is equal to the load current (on the assumption that we have a linear load). Thus, voltage from inverter 1 across the transformer is zero and power transmission is also zero. Also under this condition, inverter 2 is idling since its inverter output voltage is equal to mains voltage. Inverter 2 will supply any reactive or harmonic current from the load. In Fig. 5.14b, the mains voltage is at −15 percent and the fully regulated power to the load (±1 percent V) is obtained through inverter 2 and thus from inverter 1.

Figure 5.14c illustrates the system reaction to a +15 percent overvoltage from the mains supply. In this case inverters 1 and 2 absorb the abnormal mains voltage condition.

In Fig. 5.14d battery recharging is occurring. It is assumed that 110 percent of power is required from the normal mains source (i.e., 100 percent

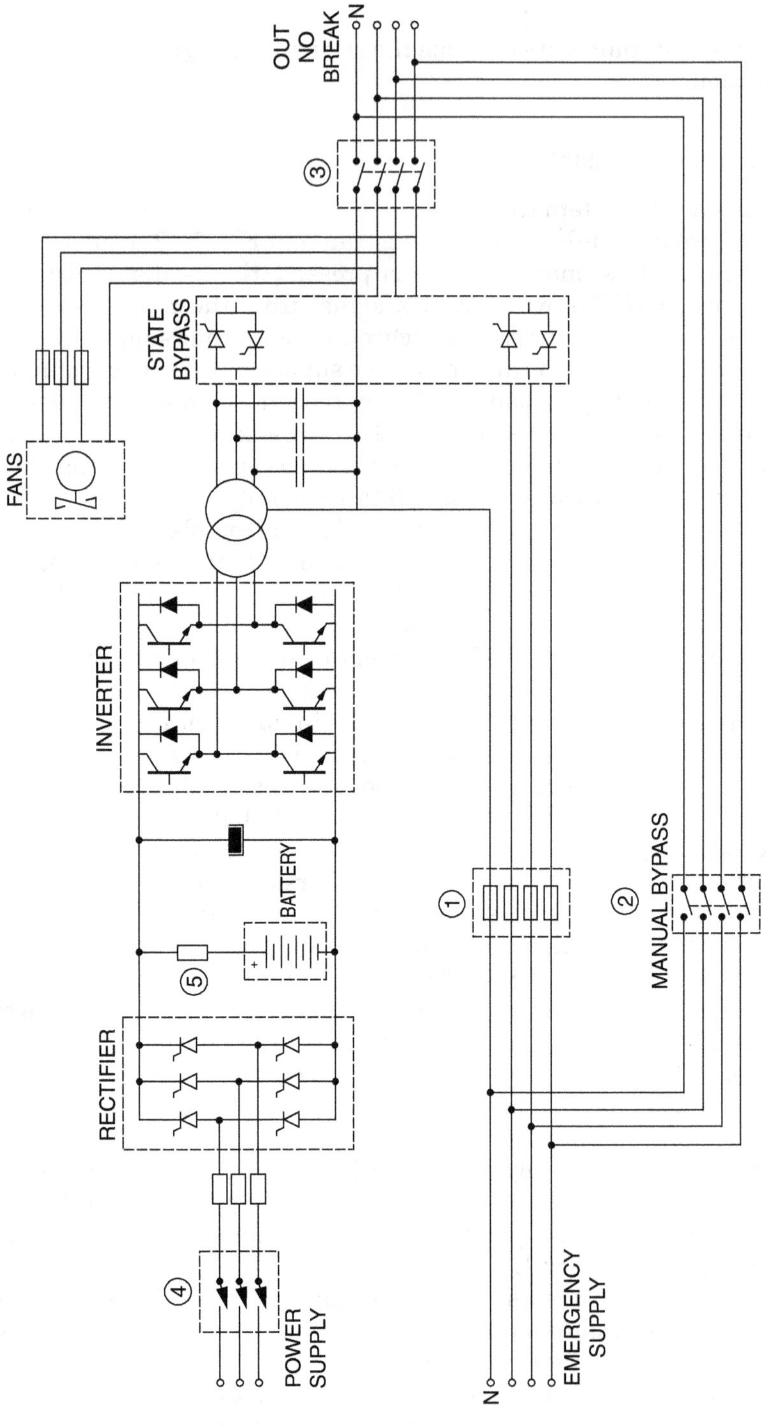

Figure 5.13

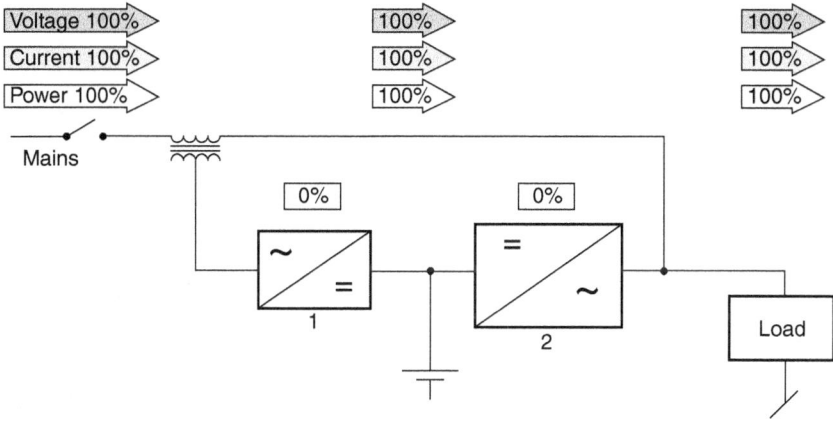

(a)

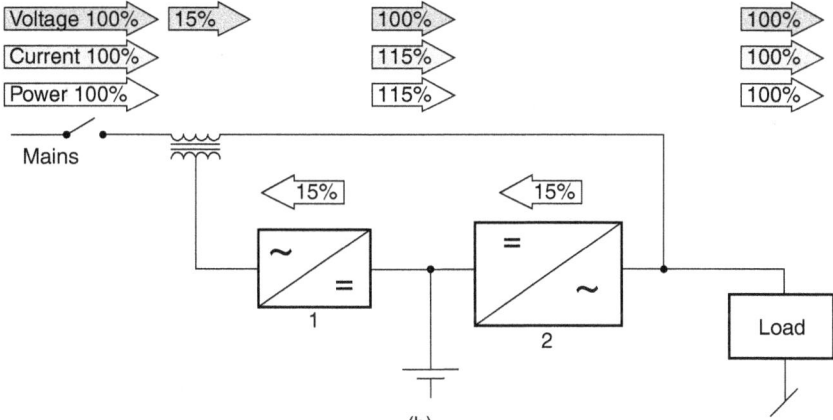

(b)

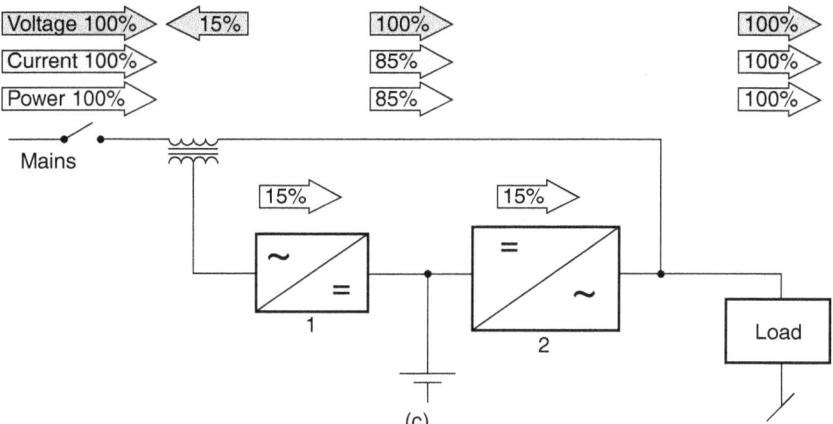

(c)

Figure 5.14

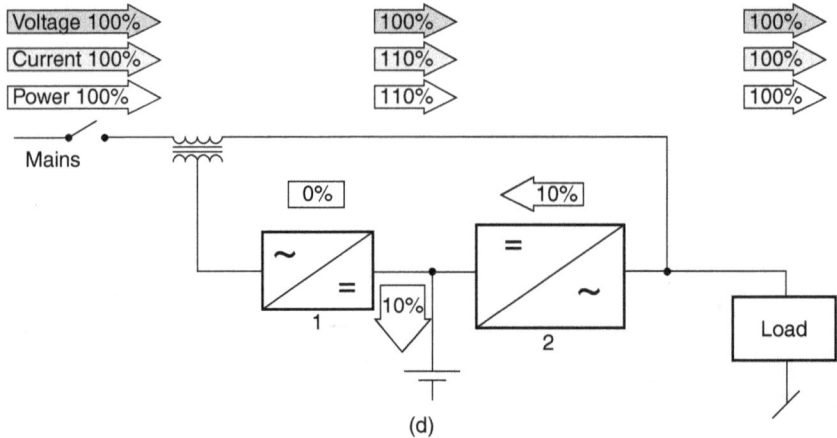

(d)

Figure 5.14

to the load and 10 percent for battery recharging). Under this condition the battery recharging is a feedback of power from inverter 2 to the battery.

It should be explained that inverter 2 is synchronized to the mains supply and acts as the voltage regulator to the load. Inverter 1 compensates for variations in power factor and, although the power for battery charging is fed from Inverter 2, the control of this function is governed by inverter 1. Additionally, inverter 1 compensates for any waveform variation between mains and the load.

Claimed advantages for this system are low harmonics induced onto the mains supply and high efficiency. The high efficiency at full load is certainly superior to a normal double conversion system but, with most UPS systems where a load factor of 70 to 80 percent is expected, there is little difference in overall system efficiency.

An example of a passive standby UPS module, wherein the load is normally fed via a switch/filter/conditioner, is shown in Fig. 5.15. Simultaneously, the battery is in charging state. On loss of mains, the load is fed from battery/inverter/filter.

Such units are normally to be seen where ratings are low, in other words, 2 kVA and below. Advantages of the module are low cost and compact, lightweight. Care should be taken to establish the qualities of the subassembly containing the switch/filter/conditioner. The switch may be a mechanical device which may not be sensitive enough for computer loads. Clearly the quality of filtering and conditioning of raw mains should also be reviewed. In addition, the system depends on mains frequency and there is no true isolation.

Line interactive UPS is shown in Fig. 5.16. Under normal operation the load is fed directly from mains with some voltage conditioning being provided by the inverter, but there is no isolation from mains. Frequency is dependent on mains supply and voltage conditioning is clearly limited. The module is competitively priced but restricted to lower rating systems.

There are many circuits similar in respect to the above basic modules described, and it is difficult to establish all the properties of UPS modules particularly at low ratings.

There is a trend dictated by the market to reduce size and weight, retaining efficiency, and above all reducing unit cost. This is particularly critical at lower ratings as one colleague said we are entering the "drip dry zone" (no-iron!). Transformers, so long part of the circuit, are considered heavy and expensive compared to alternative solutions involving simple solid-state components. Thus, circuits are appearing at ratings up to 10 kVA as shown in Fig. 5.17. This circuit allows wide

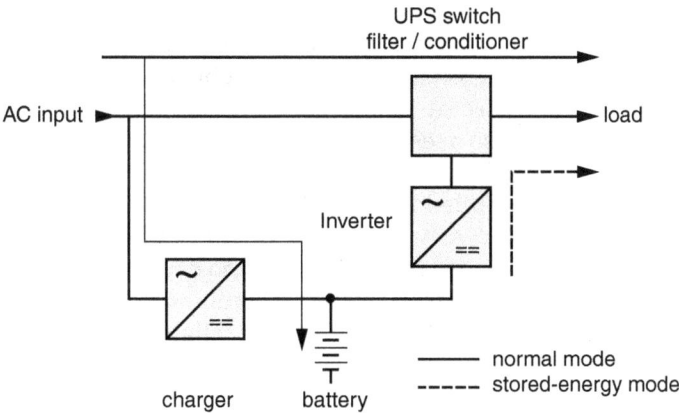

Figure 5.15

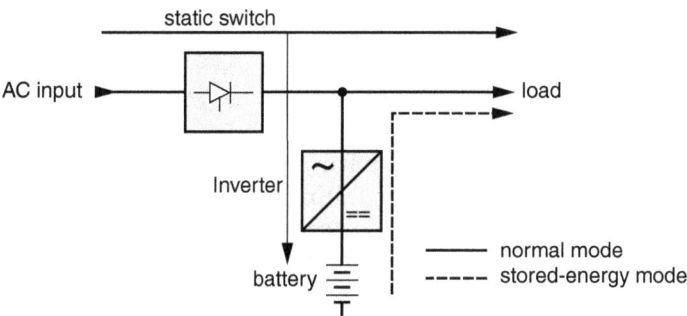

Figure 5.16

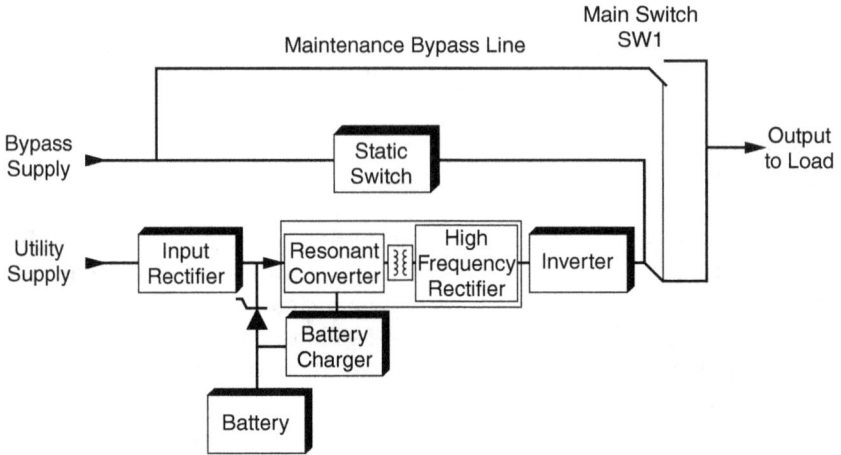

Figure 5.17

input-voltage tolerance. The resonant converter feeding the charger and the shown transformer are smaller in size than previous circuits.

Figure 5.18 illustrates a circuit where there is no large current-carrying transformer and it also uses a low-voltage battery, thus reducing considerably the cost of the battery and weight and dimensions. Note that these circuits are applied to low-rating UPS units only.

Reliability of systems has clearly improved with development and experience. Figures are hard to ascertain. There is no doubt that a single module, on-line or double-conversion type, should achieve a mean time between failure of 260,000 h. This assumes a reliable mains supply as will be met with in Europe and the United States.

Operating modules in parallel redundancy can clearly improve these figures (see Fig. 5.19). Each module consists of rectifier charger, battery, inverter, and static switch. The rating of the system is $n - 1$, where n is the number of modules in the system. Thus, failure of one module still allows full load to be maintained. In addition, static switches are used for each module. A failure of two modules results in mains supplying the load.

Synchronism of module outputs is achieved in various ways, either from a central master oscillator with an auxiliary or from each module having its own natural frequency and the modules being interconnected so the module with the highest natural frequency acts as master. Clearly in the event of the master failing, the next available set with comparative high frequency assumes control.

Paralleling modules needs care, and much development work has occurred to ensure that faulty modules do not affect the continuous safe

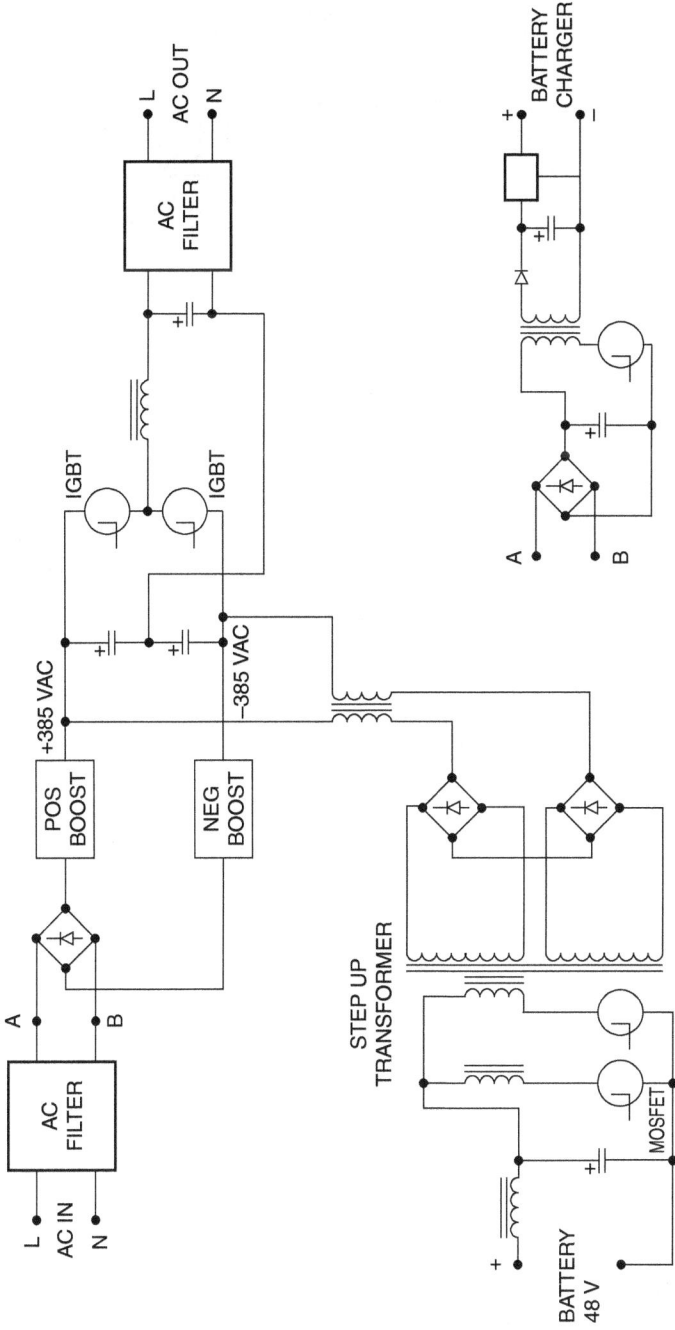

Figure 5.18

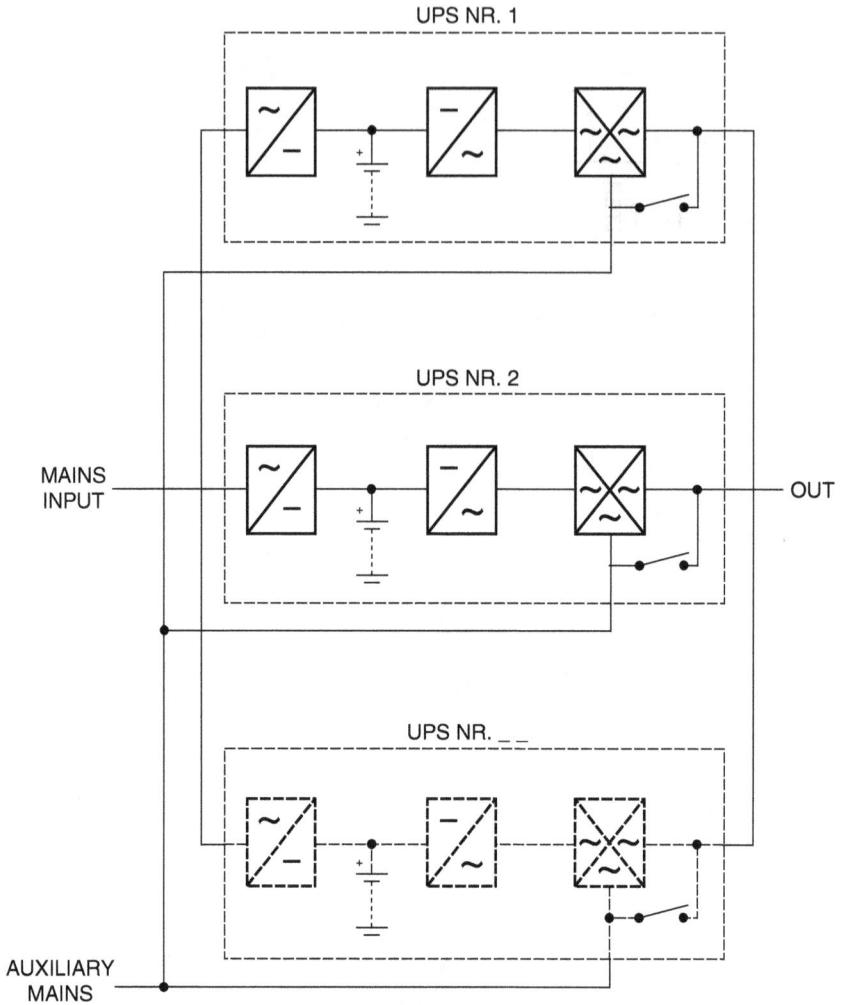

Figure 5.19

operation of the system. Clearly, under certain conditions, a heavy fault current could conceivably occur, and in practice the high-speed static switches shown in Fig. 5.17 will isolate the faulty module. Sensing the fault is of paramount concern and one chosen way is to use a voting system to ascertain the faulty module. In this system all modules send a constant stream of signals to a common control PCB, which in turn ensures correct operation and isolation of the faulty module. In many designs these interconnecting signal control cables are duplicated to ensure guaranteed operation.

The use of a common battery is not considered good practice. There is the possibility of circulating currents due to slight variations in charger performance, but more significant is the effect on system reliability.

Development of more elaborate systems is illustrated in Fig. 5.20. In panel a the three modules can be operated with either mains supply 1 or 2 and loads can be mechanically switched between the two alternative bus bars. In panel b there are two separate parallel systems, each with its own alternate mains supply, and the loads can choose to operate with either system. The load transfer module is shown in Fig. 5.21. This design ensures that there will be no break in supply to the load if either of the two parallel redundant systems fails.

Monitoring

Monitoring of UPS units has advanced considerably and, by using telephone lines and electronic mailing systems, copious amounts of information and control are now available remotely. Systems now can be maintained from a remote maintenance area and, indeed, if required, controlled and in certain instances first aid repair can be implemented.

A complete log of all events is available. This will include any voltage/power variations to the system, usage of the battery, and usage of a standby prime mover in the event of a prolonged outage. In multimodule systems such full redundant systems may also be connected and full details are made available. Some indication of the enormous amount of information available will be gleaned by examining Figs. 5.22 and 5.23.

The availability of such monitoring techniques has had a profound effect on operational requirements of UPS facilities, gone are the days which I can remember when one received a telephone call during unsocial hours from a member of security staff requesting what action should be taken due to flashing lights on UPS systems when the equipment was correctly responding to a power outage!

Maintenance on site is now virtually eliminated and fully qualified staff are immediately aware of any events on site. Thus, costs are reduced and any remedial response is immediate.

Bibliography

BS 50091-1-1993 or EN 50091-1

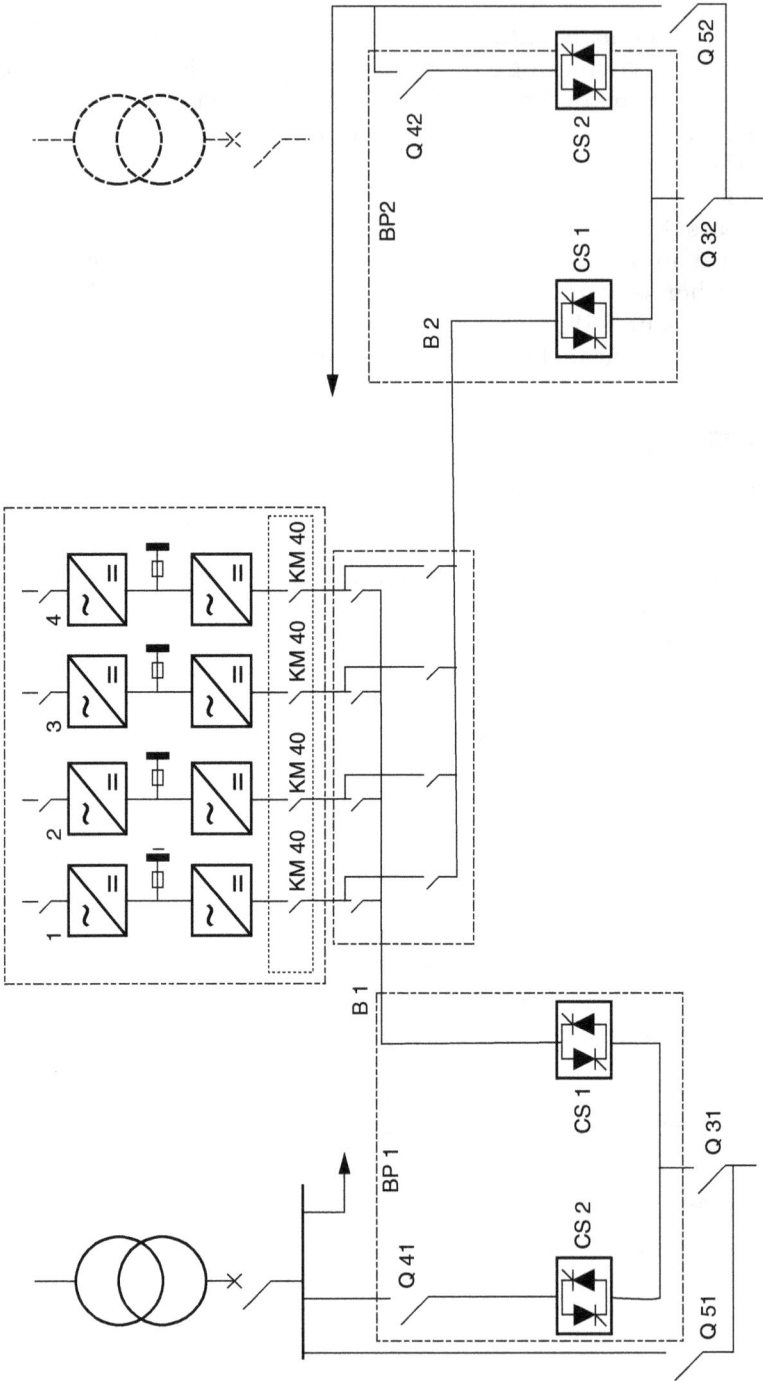

Figure 5.20

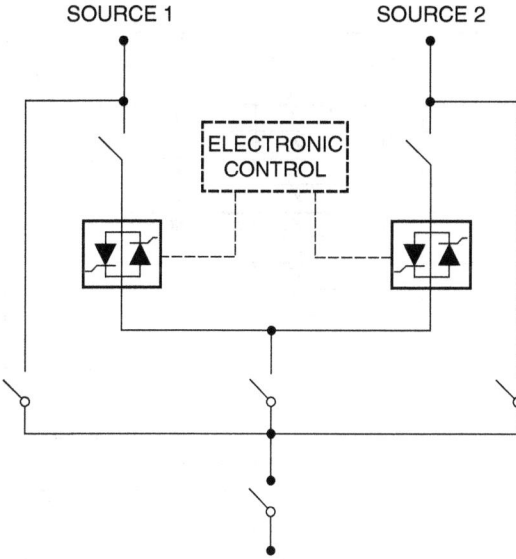

Figure 5.21

Features Available by Connection Type

Category	Feature	Local		Network	
		Contact Closure	Serial UPS	MultiLink Basic Notification Source	SNMP
Overview Data (Overview tab with UPS selected)	Graphs	No	✓	No	✓
	Parameters	No	✓	No	✓
	Info	No	✓	No	✓
	States	✓	✓	✓	✓
Data Categories (Overview tab with UPS selected and expanded)	Battery Group	No	✓	No	✓
	Output Group	No	✓	No	✓
	Bypass Group	No	✓	No	✓
	Input Group	No	✓	No	✓
	Internals Group	No	✓	No	✓
	Alarm Group	No	✓	No	✓
UPS Control (Overview tab right-click on UPS for pop-up menu)	Diagnostics – Test Battery	No	✓	No	✓
	Output Control – Turn UPS Output Off/On	*	✓	No	✓
	UPS Audible Alarm – Silence	No	✓	No	✓
UPS Events (Event Configuration tab with UPS selected)	UPS Operating On Battery	✓	✓	✓	✓
	Low Battery	✓	✓	✓	✓
	UPS Output Returned From Battery	✓	✓	✓	✓
	Other Events	None	All (see Table 11)	None	All (see Table 11)
Event Actions (Event Configuration tab with UPS selected)	Notify	✓	✓	✓	✓
	E-mail	✓	✓	✓	✓
	Page	✓	✓	✓	✓
	Shutdown	✓	✓	✓	✓
	Command	✓	✓	✓	✓
	Log	✓	✓	✓	✓
	Silence Alarm	No	✓	No	✓
Monitoring Device Events (Event Configuration tab with My Event Actions [or monitoring device] selected)	System Shutdown Commencing	✓	✓	✓	✓
	UPS Monitor Software Started	✓	✓	✓	✓
	UPS Monitor Software Stopped	✓	✓	✓	✓

Figure 5.22

Configuring the Software

Function	Access Method	For information, see:
Save changes to event actions	MultiLink menu	9.1.1 – MultiLink Menu — Save Event Configuration
Start or stop MultiLink service		9.1.2 – MultiLink Menu — Service
Exit Viewer		9.1.3 – MultiLink Menu — Exit
Rename a UPS	Right-click on UPS for Properties (Overview tab)	9.2 – Renaming a UPS
Test battery; turn UPS output on or off; temporarily silence an alarm	Right-click on UPS for UPS Control (Overview tab)	9.3.2 – Review the UPS Control Options
Enable alerts; change UPS connection type or communication ports; change status polling area	Right-click on UPS for Properties (Overview tab)	9.3.3 – Properties — Setup Tab
Change UPS settings: enable audible alarm, battery testing and auto restart		9.3.4 – Properties — UPS Settings Tab
Add MultiLink Network Shutdown Clients		9.3.5 – Properties — Clients Tab
Configure data logging features		9.3.6 – Properties — Data Logging Tab
Add a local UPS	Configure Menu	9.4.3 – Configure Menu — Edit My Device List
Add a network MultiLink installation		9.4.4 – Configure Menu — Edit Network Device List
Enable and configure modem		9.4.5 – Configure Menu — Modem
Change password or permission for remote configuration		9.4.6 – Configure Menu — Local Password
Install an upgrade license		9.4.7 – Configure Menu — Upgrade License
Setup responses to UPS-related events	Event Configuration tab	9.5 – Configuring — Event Configuration tab
Simulate a UPS-related event to test configuration	Right-click on event for Test Event (Event Configuration tab)	9.6 – Testing Responses to Events
Temporarily disable a UPS-related event	Right-click on event for Disable (Event Configuration tab)	9.7 – Disabling an Event
Temporarily disable an event Action	Right-click on Action for Disable (Event Configuration tab)	9.8 – Disabling an Action

Figure 5.23

6

Rotary UPS Systems

Definitions

This chapter covers rotating sets capable of providing a continuous, well-regulated power supply to the load irrespective of the normal mains supply. There would appear to be various methods available to create such a system, and they are probably best discussed under two headings: rotating transformer systems and close coupled diesel/clutch/generator sets.

Background

A very early system which one could hardly call a UPS system merely consisted of motor/heavy flywheel/generator system. At this stage engineers were trying to bridge small power transients. The design was not very successful, the system would provide an enormous amount of power. In most designs the output voltage held up for 5 to 10 s but within 1 s the output frequency was out of tolerance. Note that this idea of utilizing kinetic energy has been given a new lease on life due to modern electronic circuitry (see Chap. 8).

For many years prior to the advent of the static system the circuit illustrated in Fig. 6.1 was in use. It consisted of a rectifier and battery charger supporting the load and also charging a battery. The dc supply was used to drive a motor alternator set connected to the load. The flywheel was present to provide kinetic energy when the system switched from mains supply to battery supply. At this instant the battery would go from float voltage instantaneously to discharge voltage a drop of some 15 to 17 percent. The regulating device dynamic performance was too poor to cope with this, and the kinetic energy from the flywheel ensured a reasonable output performance.

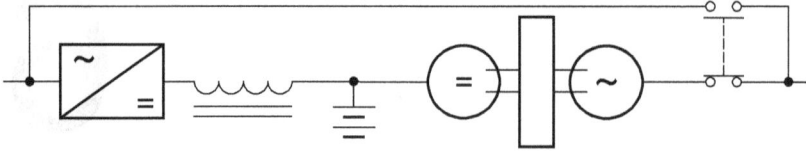

DC Motor driven Rotary Output Stage with Battery Energy Store

Figure 6.1   Typical basic dc machine rotary UPS.

Drawbacks to the system were clearly the brushes' requiring maintenance and the systems's low efficiency of approximately 80 to 82 percent.

The advent of the static system with its higher efficiency, improved
performance, and promised low maintenance sounded the death knell
of this circuit. But not without a fight!! There was a brief period when
the reliability of the static UPS was called into question and coupled
with this was the problem of site maintenance. Most maintenance staff
at that time fully understood the rotary system, and advanced elec-
tronics caused a service problem. Provided that the rotary system was
on a fairly constant load, the brush problem was greatly attenuated.

Modern designs are highly efficient and tend to be applied for higher
rated systems.

Rotating Transformer Systems

The rotary transformer systems produced by Anton Piller are offered in
a variety of circuits The basic rotary system is shown in Fig. 6.2. Under
normal mains operation the load is fed via the static switch and rotating transformer directly. Simultaneously, the rectifier charge ensures
the battery is in the float condition and the inverter is in operation but
not providing power to the rotating transformer. The rotating transformer (Uniblock) has alternate primary and secondary transformer
windings in the stator. The rotor provides regulation of both voltage and
frequency. The rotary transformer clearly provides galvanic isolation.

On loss of mains supply the rotary transformer obtains power from
the inverter/battery system, and the static switch isolates the failed
incoming supply. The system is highly efficient, with a claimed figure
of 95 percent, and no waveform distortion is induced onto the incoming
supply.

The output short circuit current is high (14 × In).

In lieu of a battery, a flywheel kinetic energy source maybe used as
discussed in Chap. 8.

An alternative system is to use both Uniblock windings to supply
critical and essential loads via a directly coupled diesel engine as
shown in Fig. 6.3. The Uniblock windings are independent in such a

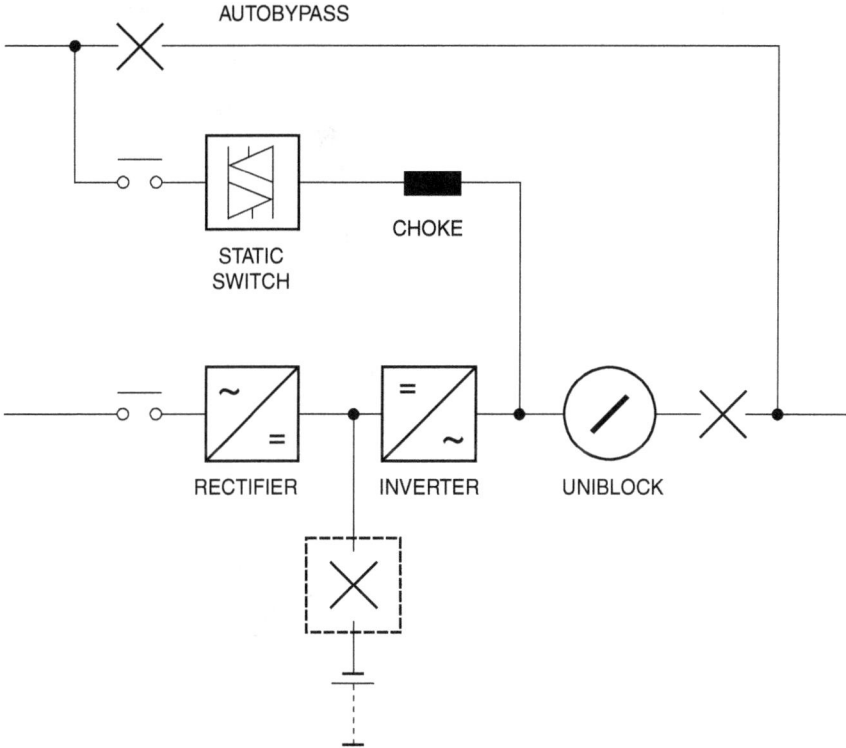

Figure 6.2   UPS employing rotating transformer system.

system, neither affecting the operation of the other. Thus, the critical load will give the same performance as shown in Fig. 6.2.

The Uniblock system in Fig. 6.2 is also available using the water-cooling system of the building thus reducing the system size and air-conditioning load and allowing the system to be situated in some unusual areas.

Parallel redundancy is available and a typical circuit is illustrated in Fig. 6.4—in this case with static switches between the various paralleling systems adjacent to the loads.

Generator/Clutch/Machine

There are available a number of rotary systems employing diesel/clutch configurations and we detail their operation below.

Our first example of close coupled diesel/clutch/generator is provided by the system illustrated in Fig. 6.5 as produced by Eurodiesels Ltd. Under normal mains operation power fed into the machine rotates the synchronous machine at 1125 rpm. Coupled to this machine is the field

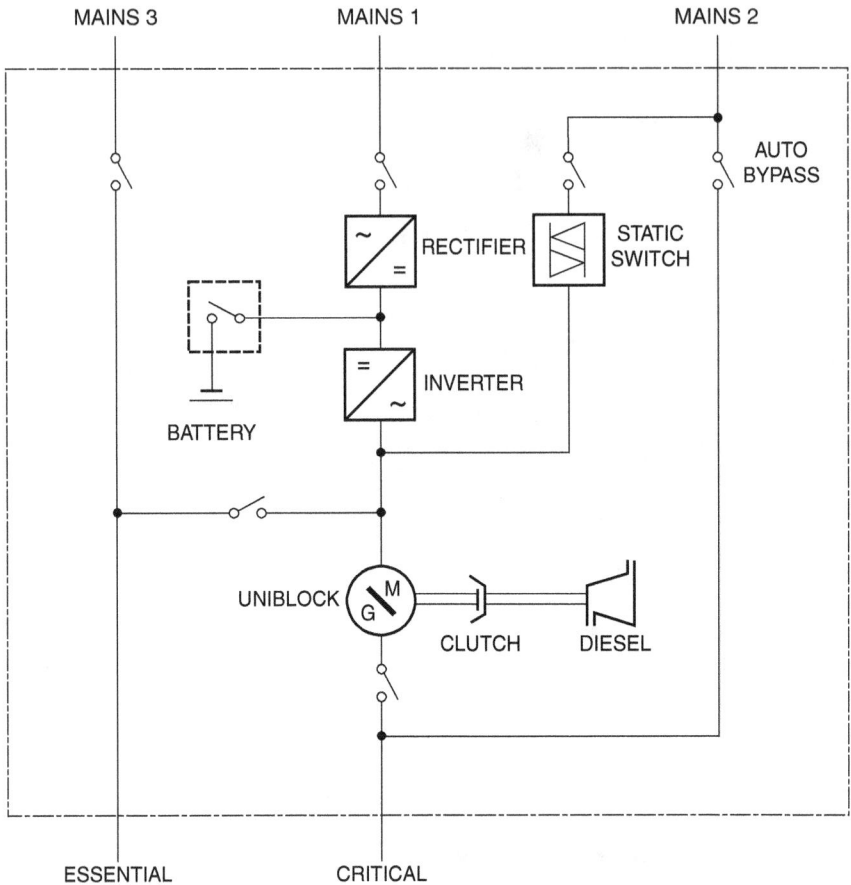

Figure 6.3 Rotating transformer system with directly coupled diesel engine.

winding of an asynchronous motor, clearly rotating at the same speed. The squirrel cage rotor designated in Fig. 6.5 as the Accu rotor turns at speeds of 1500 rpm relative to the speed of the field winding. Thus, the Accu rotor is in reality turning at approximately 2600 rpm (allowing for slippage). The speeds indicated above are for a system operating on a 50-Hz supply. Such sets will operate on a 60-Hz supply, the speeds will increase pro rata.

On loss of mains supply the incoming breaker is opened and kinetic energy from the high-speed Accu rotor is utilized to ensure that the main rotor is kept within tolerance. On opening the mains input breaker, the diesel is electrically started and within 1 to 2 s the generator has reached a speed of some 200 to 300 rpm. At this point the electromagnetic clutch

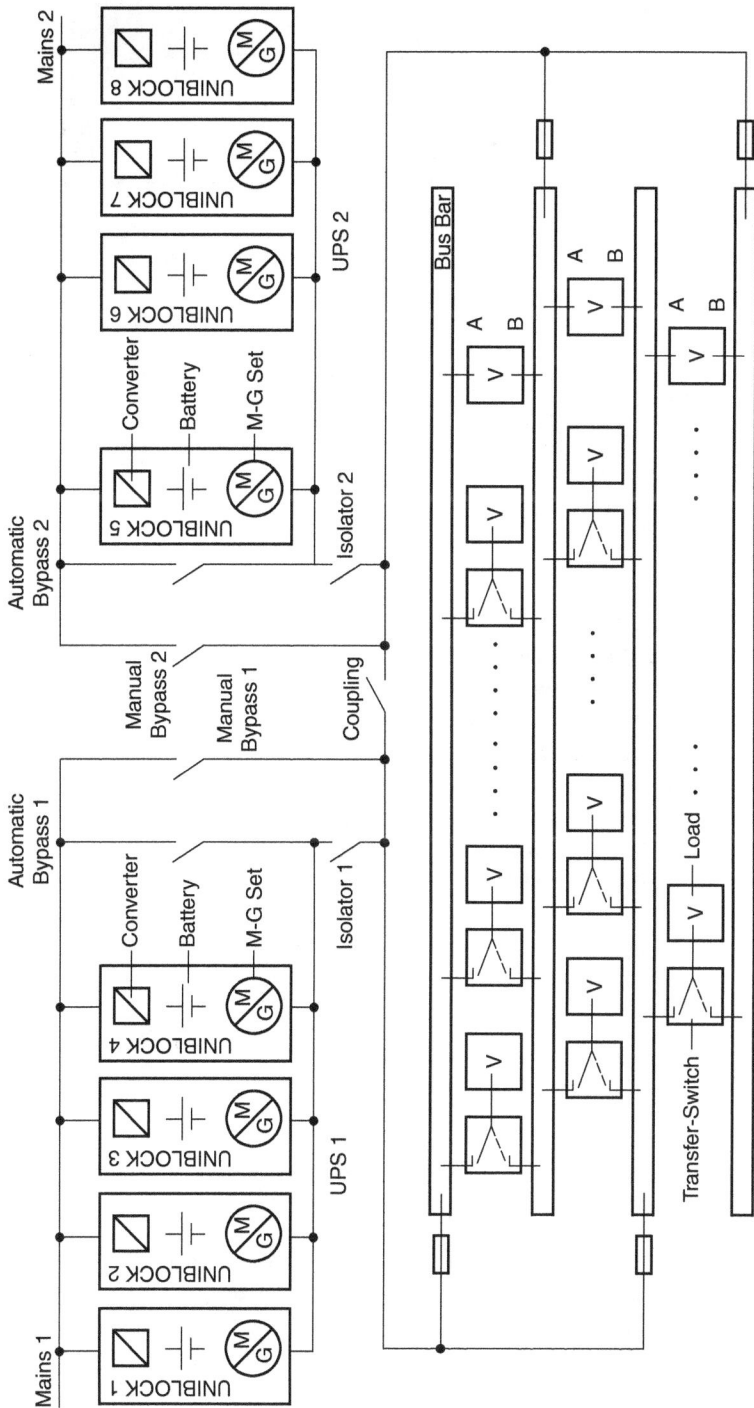

Figure 6.4 Parallel redundant system with additional static transfer switches.

173

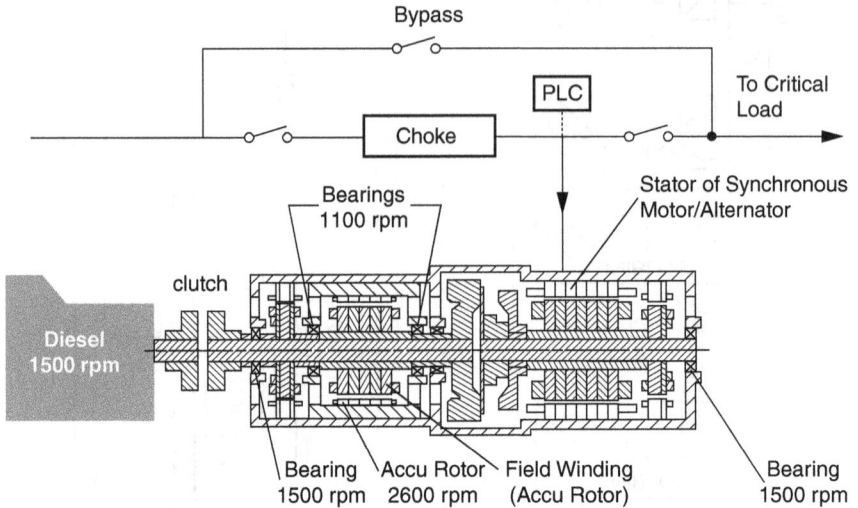

Figure 6.5 Close coupled diesel/clutch/generator from Eurodiesels Ltd.

is engaged and the kinetic energy from the high-speed rotor is also used briefly to bring the generator up to speed.

In more detail the system will ride through microcuts up to 300 ms. Beyond this time the synchronous rotor functions as an alternator and instantaneously feeds the critical load without voltage variation, and its nominal speed decreases. Speed decrease will clearly affect the output frequency, and at 49.8 Hz the inductive coupling between the two rotors is activated (in less than 10 ms). As a result, the main rotor accelerates and the Accu rotor slows down.

At point 0 in Fig. 6.6 the engine is started electrically, and 1 s after point 0 the electromagnetic clutch is activated and the mechanical time to close it smoothly is 0.5 s.

Thus 1.5 s after mains failure there is a mechanical connection between the engine running at about 400 rpm and the main rotor running between 1494 and 1500 rpm. Clearly, there is clutch slippage at this point and the main rotor continues to receive more energy from the Accu rotor. This in turn accelerates the diesel which attains 1500 rpm in ±1 s.

The electronic speed regulator of the engine adjusted at ±1506 rpm (50.2 Hz) will become the master frequency regulator and the engine will ramp up the load and continue to drive the stato–alternator which feeds the critical load. On achieving frequency stability the signal is given to increase speed of the Accu rotor to its nominal 2500 rpm. Upon restoration of the mains supply and synchronization being achieved the mains breaker is closed.

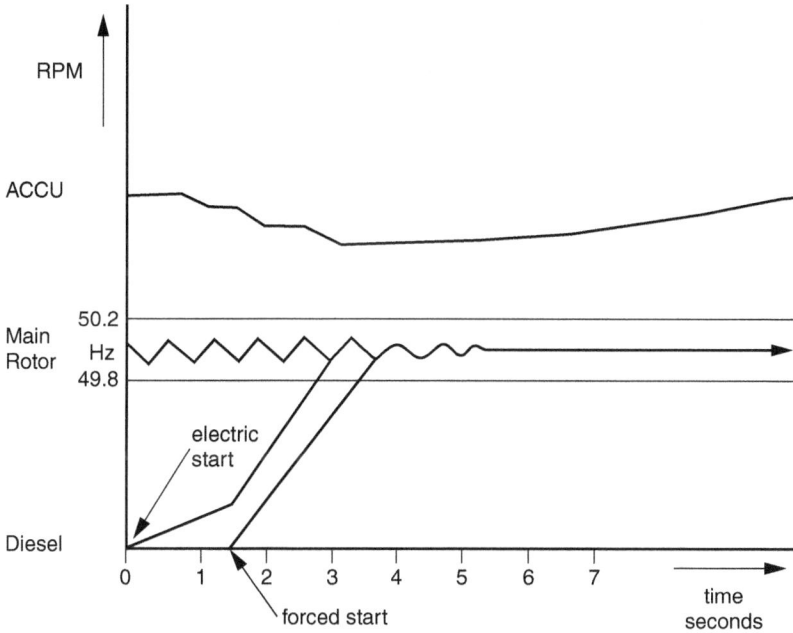

Figure 6.6 Transfer graph—failed mains to diesel operation.

Figure 6.5 shows a choke in the line between the normal mains sup-ply and the load. This has a number of functions. The inductance acts as a harmonic filter. Thus, the system complies with the low harmon-ics called for in both the United States and Europe (approximately 5 percent). It limits full load current from mains supply to 4 times maximum load figure. The inductance working with the synchronous motor/alternator acts as a voltage-regulating device in mains/load operation.

Figure 6.7 illustrates the waveform to be expected onto the mains sup-ply and the waveform onto the load when operating with a load high in harmonics (e.g., a computer load with a high switch mode power system).

The performance of the system shows an efficiency varying between 93 and 96.4 percent, voltage regulation of nominally ±1 percent, a load switch of 100% resulting in a ±5 percent performance, and a dynamic recovery to ±1 percent within 100 ms.

Frequency regulation is ±0.2 percent and a 50 percent load switch gives a ±1 percent Hz figure. The short circuit figure is 20 × In from machine to load.

Claimed advantages for the system are high efficiency (93–96.4 per-cent), a claimed mean time between failure of 50 years, comparative smaller required floor area, and the enhanced ability to clear subcircuit

Input (Upstream of KS) Waveform L_1, L_2, L_3

Site: 2 MVA at ING Barings

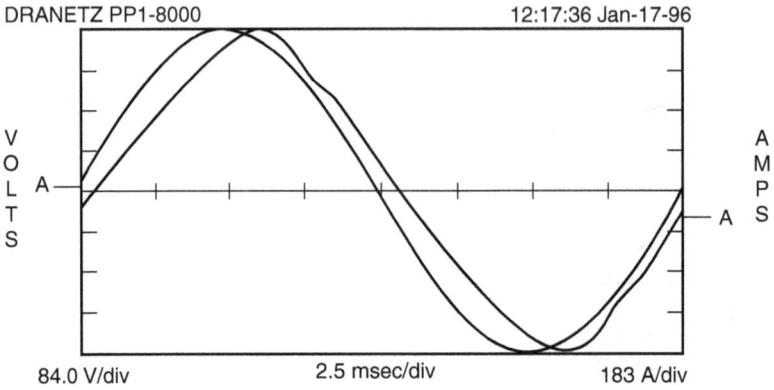

DRANETZ PP1-8000   12:17:36 Jan-17-96

84.0 V/div   2.5 msec/div   183 A/div

OBSERVATION
> The voltage and current waveforms are almost in phase.
> Excellent power factor correction.
> The voltage and current waveforms are sinusoidal giving good reflection onto mains.

Output (Downstream of KS) Waveform L_1, L_2, L_3

Site: 2 MVA at ING Barings

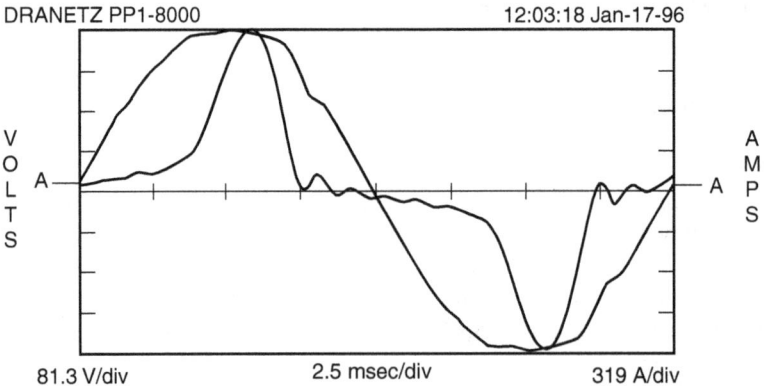

DRANETZ PP1-8000   12:03:18 Jan-17-96

81.3 V/div   2.5 msec/div   319 A/div

OBSERVATION
> The rich harmonic current load creates poor current waveform.
> However output voltage waveform from No-Break KS remains sinusoidal.

Figure 6.7   Illustration of attenuation of load waveform distortion.

faults. It should be remembered that rotary sets do require more maintenance, but this type of circuit does not require battery systems.

The system may also be used to supply both critical and noncritical loads, simply by increasing the rating of the diesel engine, and making switch gear changes. Figure 6.8 illustrates this.

The noncritical load is available within 10 s of mains failure, the use of this system has no effect on the quality of the power to the critical load.

There are various parallel/redundant circuits available (Fig. 6.9). In this system under normal operation all units are in synchronism and phase working from mains supply, under loss of mains supply the diesels will automatically synchronize within 0.5 s. It should also be noted that the static switches will operate up to 15 degrees out of phase.

Other parallel circuits are available with and without the option of supplying critical and noncritical loads. Figure 6.10 illustrates a complex dual parallel redundant system.

The second alternative we examine (being second for examination does not in any way imply any system defect!) is produced by Hitek, previously known as Holec. This system is lighter and somewhat smaller, runs at a slightly higher speed, and differs in the induction connection and the clutch methodology employed.

Basically, mains supplies the load via the reactor and the three-phase generator, illustrated in Fig. 6.11, which then act as a filter inhibiting any mains problem affecting the load. Under this condition the generator acts as a motor and drives the outer rotor of the induction coupling

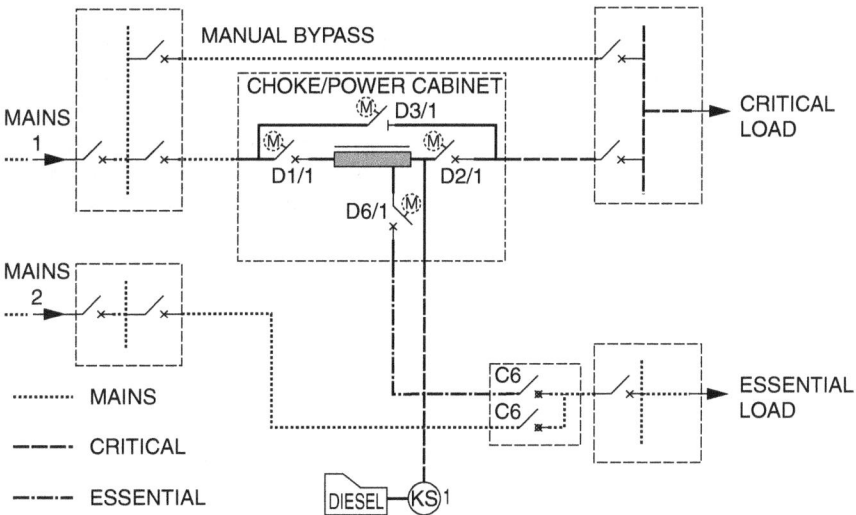

Figure 6.8   System feeding both critical and noncritical loads.

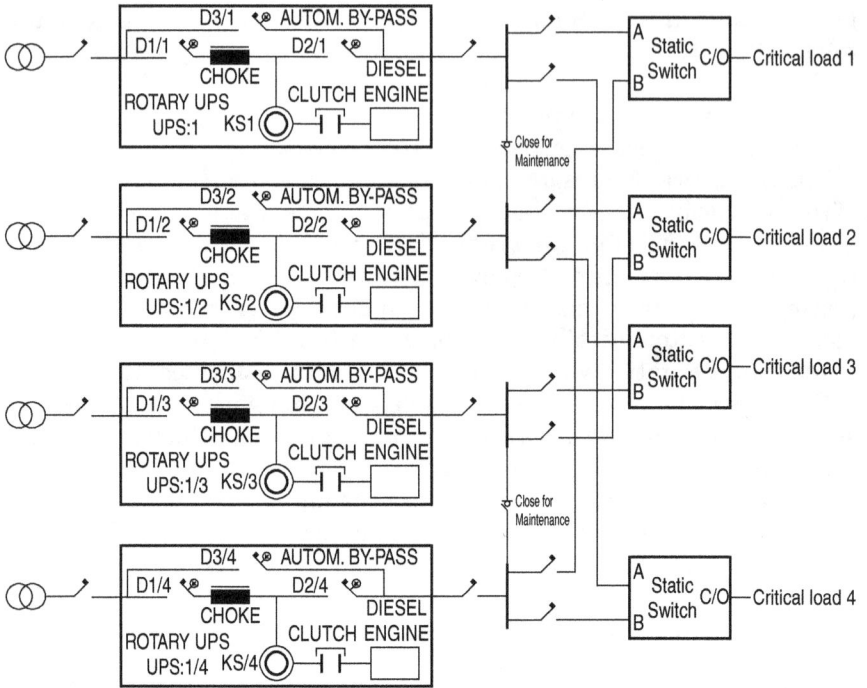

Figure 6.9 Parallel redundant circuit.

at 1500 to 1800 rpm (speed is dependent on mains frequency 50/60Hz). There is a two-pole three-phase winding in the outer rotor and this, on excitation, ensures that the inner rotor has a speed of 3000 to 3600 rpm. A free wheel clutch isolates the outer rotor from the, at this juncture, quiescent diesel engine.

On loss of mains the circuit breaker Q1 is opened and the dc windings on the induction coupling are energized and kinetic energy from the high-speed inner winding ensures that the outer rotor maintains its speed. Simultaneously, the diesel ramps up to 1500 to 1800 rpm within 2 s or less and the free wheel clutch engages automatically. Within 10 s the diesel assumes supplying full power to the system

In the gap between point zero of load transfer and 10 s, power is obtained by a combination of kinetic energy from the inner rotor and the diesel engine. This is sufficient to retain a high quality voltage and frequency regulation to the load. In a short time period the inner rotor is accelerated to its nominal operational speed of 3000 to 3600 rpm. It should be noted that the nominal speeds indicated are the speeds relative to the speed of the outer induction winding and, thus, are in fact operating at 4500 to 5400 rpm.

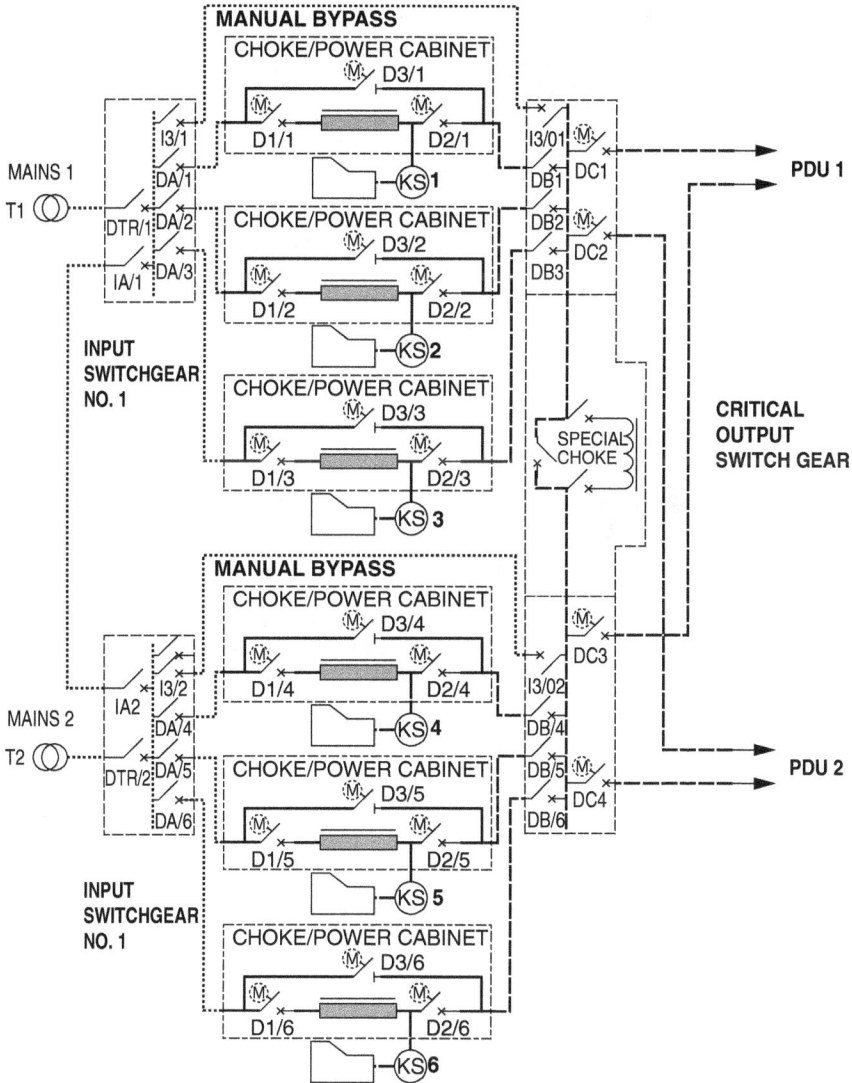

Figure 6.10 Complex parallel redundant configuration.

On restoration of a stabilized mains supply, with synchronism being achieved, circuit breaker Q1 is closed and the diesel engine is allowed to ramp down to 1450 to 1750 rpm, at which time the free wheeling clutch is disengaged automatically.

The system will accept ±10 percent voltage and ±1 percent frequency. Outside these tolerances the system will revert to diesel operation. To inhibit the problem of short transients initiating the diesel

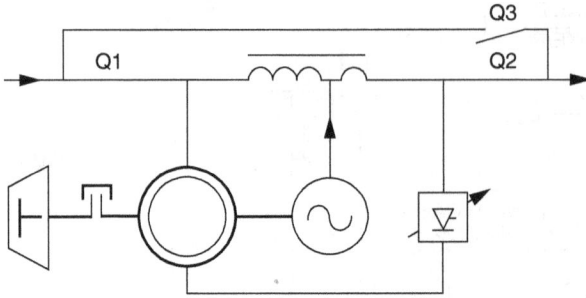

Figure 6.11   Hitek basic circuit configuration.

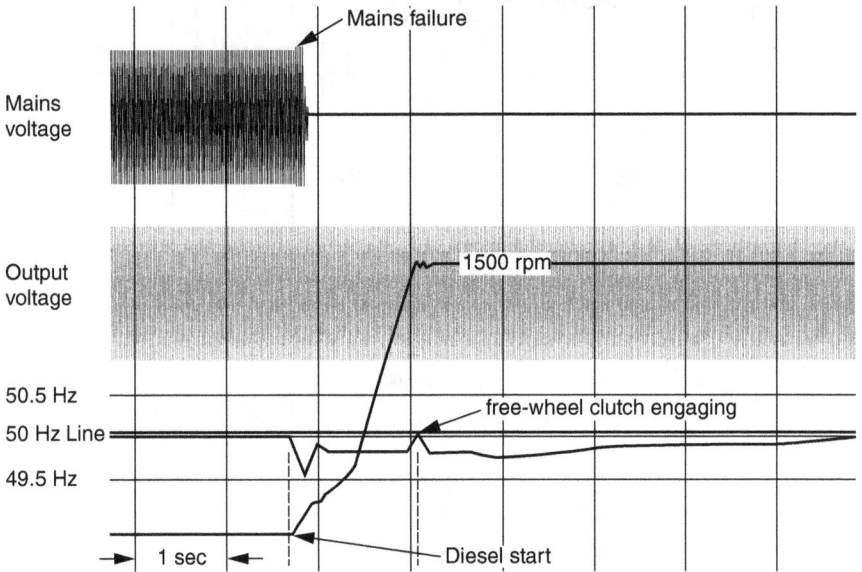

Figure 6.12   Effect of loss of mains at full load.

start-up procedure, a delay of approximately 0.5 s is utilized. The delay can be even longer than this as the system has a sensitive voltage monitoring system which will differentiate between a deep voltage excursion and minor ones which in some instance will be discounted. Overall system efficiency may be as high as 97 percent.

Figure 6.12 illustrates the varying output voltage and frequency of a changeover at full output power.

The system may also be used to feed noncritical loads. In Fig. 6.13, Fig. 1 illustrates operation with mains present, noncritical load being fed directly from mains supply. Fig. 2 illustrates loss of mains and the critical load being fed without interruption. At this stage diesel has not achieved full running of the system and noncritical load is not fed. With diesel in full operation both critical and noncritical loads are fed, Fig 3 of Fig. 6.13. Obviously, with such a system the diesel has to be rated for the full critical and noncritical loads.

Parallel redundancy can be achieved easily in a variety of ways, parallel operation of both critical and noncritical loads being possible. Also, static switches may be used in conjunction with system bypass circuits to provide a guaranteed noninterrupted load if so required. Figure 6.14 shows the waveform of the output from a parallel redundant 1600-kVA system with 800-kVA load rupturing a 400-amp motor start fuse. The photo in Fig. 6.15 is of a parallel redundant system manufactured by Hitec, installed at a large banking computer center.

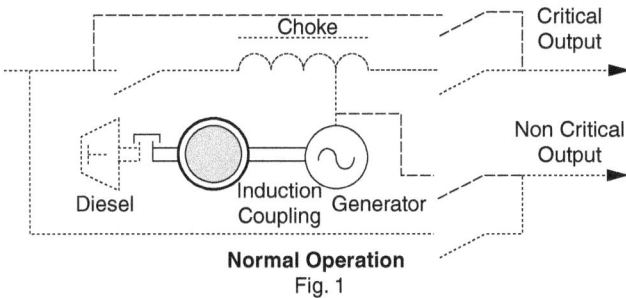

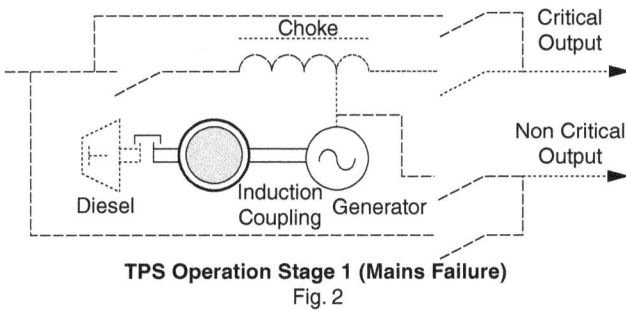

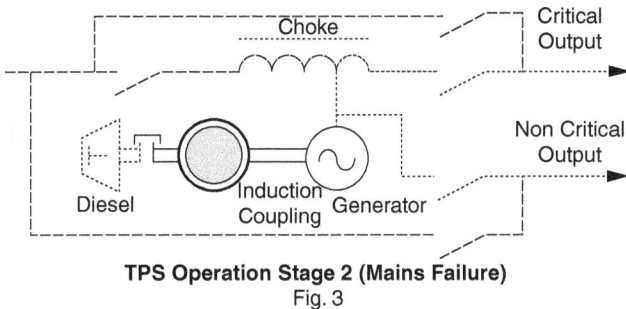

Normal Operation
Fig. 1

TPS Operation Stage 1 (Mains Failure)
Fig. 2

TPS Operation Stage 2 (Mains Failure)
Fig. 3

Figure 6.13

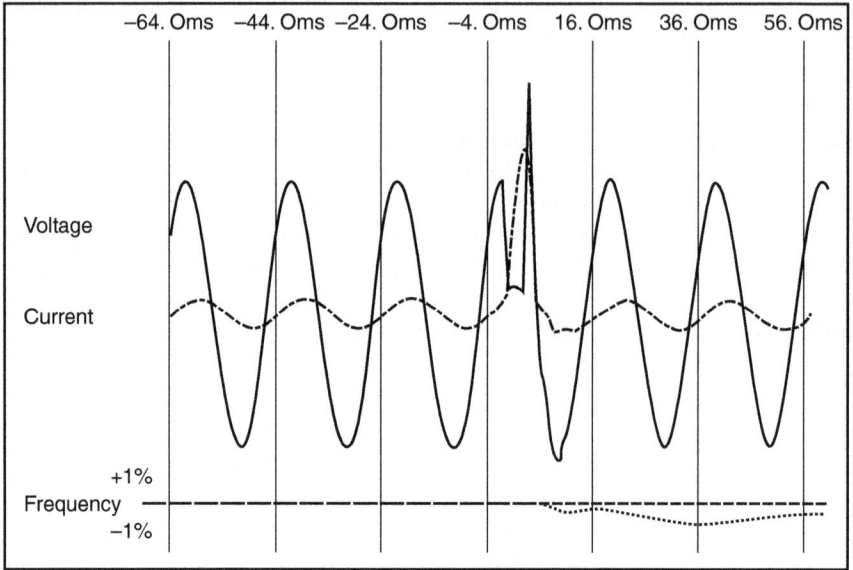

Figure 6.14 The effect of rupturing a 400A motor start fuse on a 2 × 800 kVA parallel system feeding an 800-kVA load.

Figure 6.15 Typical machine room for a large computer system.

The system consists of a remote diesel consisting of a kinetic energy store in the form of a rotating flux field inductor/flywheel rotating at 3600 rpm under normal mains operation. The bearings are assisted by magnetic levitation. A synchronous machine is coupled to an inductor acting as a filter under normal mains operation.

Figure 6.16 illustrates the basis of the system. Under normal mains operational condition the load is supplied via inductor and the synchronous machine and inductor act as a filter ensuring the load sees no spikes or voltage dips. The system, on loss of mains supply, initiates the start up of the supporting diesel. Meawhile, kinetic energy within the flywheel/flux field inductor system is ensuring that the synchronous

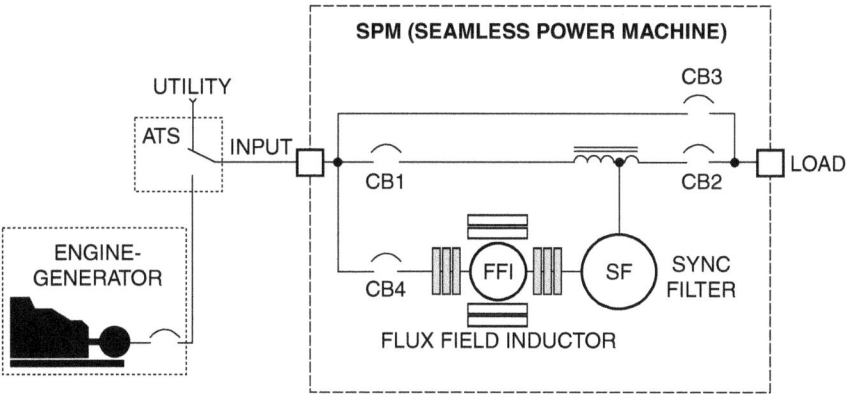

Figure 6.16   Basic circuit diagram of system as manufactured by Perfect Power Systems California.

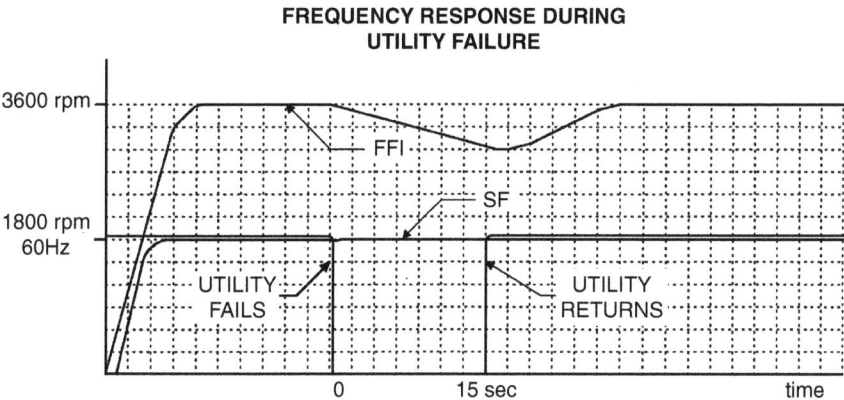

Figure 6.17   Frequency response on switch over.

machine continues to run at its allotted speed of 1800 rpm. The system ensures the load sees very little disturbance in quality of power supplied (see Fig. 6.17). This system has the advantage of a remote diesel generator set, a high efficiency of 97 percent, and being housed in a cubicle construction, noise is attenuated to reasonable levels. The manufacturer is Perfect Power Systems, USA.

Acknowledgments

The authors acknowledge the assistance given to them in preparation of this chapter by the following companies:

Anton Piller, Osterode, Germany

Hitec Power Protection, Hengelo, Netherlands

Eurodiesel, Liege, Belgium

7

Batteries

Introduction

It is said "95 percent of UPS problems are induced by battery/human error problems," a sweeping statement, nevertheless one in general vogue at the moment. True, in the early days of development of the popular sealed lead acid battery, now correctly identified as the valve regulated lead acid (VRLA) battery, reliability was not high. Since that time some 20 years ago, there have been constant developments in design resulting in a reliable performance. This chapter therefore goes into some detail on types of VRLA cells available and also covers the flooded-type cells available today.

Types of Cell

Among the variety of secondary (rechargeable) cells available, only two are in use for UPS applications. These are lead acid and nickel-cadmium batteries. Both of these are available in differing types, each type having its own characteristics and qualities affecting choice and application.

Lead Acid

Lead acid cells having a chemical process which in simplistic terms consists of electrodes constructed from lead and lead dioxide in a dilute sulphuric acid solution. If the potential is measured between the two electrodes, an approximate reading of 2 V is apparent.

Chemical Reactions/Basic Design

The chemical reaction that occurs is best illustrated as follows:

1. Reaction at the negative plate (pure lead)

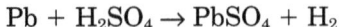

$$Pb + H_2SO_4 \rightarrow PbSO_4 + H_2$$

2. Reaction at the positive plate (lead oxide)

$$PbO_2 + H_2SO_4 \rightarrow PbSO_4 + (2OH)$$

Adding together 1 and 2 the chemical reaction for the cell is:

$$PbO_2 + Pb + 2H_2SO_4 \rightarrow 2PbSO_4 + 2H_2O$$

As the action proceeds and the discharge is complete, the plates are covered in the case of the negative plate with lead sulphate and the positive plate with a mixture of lead compounds, PbO and PbSO.

Recharging the cell involves converting the lead sulphate to lead. Thus, the reactions are now as follows:

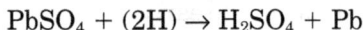

$$PbSO_4 + (2H) \rightarrow H_2SO_4 + Pb$$

The peroxide is regenerated as follows:

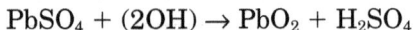

$$PbSO_4 + (2OH) \rightarrow PbO_2 + H_2SO_4$$

The complete reaction is thus expressed as:

$$2PbSO_4 + 2H_2O \rightarrow Pb + PbO_2 + 2H_2SO_4$$

To all intents and purposes the efficiency of the reaction is 75 to 85 percent, that is, the losses between discharge and recharge (Fig. 7.1).

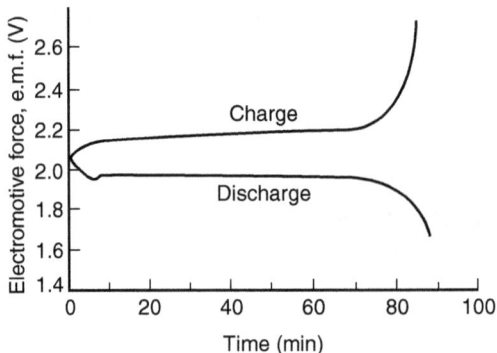

Figure 7.1

The cell, irrespective of types, shown in Fig. 7.1 consists of the following components:

Positive plate

Separators

Negative plate

Electrolyte

Container

Vents

Intercell connectors

Types of Cell: Plante/Tubular/Pasted Plate/VRLA

The design of the plates gives various qualities to the cell. The types which interest us are Plante, pasted plate, tubular, and the valve regulated lead acid (VRLA) cell. Each of the aforesaid designs has certain characteristics which may be suitable to the UPS design in question. It should also be noted that further variations in design of the types mentioned above can be expected, for example, variations in plate structure and thickness, specific gravity of the acid employed, and the alloy utilized in the plate construction.

The actual thickness of the plates determines the availability and length of the discharge; the surface area of the plates has a very significant effect on the current output.

Since most UPS applications call for a short autonomy period, the plate design is usually fairly thin. A comparison of the qualities of Plante, tubular, and VRLA cells is shown in Fig. 7.2. The figures are based on equivalent A/Hr rated plates.

The Plante cell was used extensively in the early days of UPS development, but its design and cost have relegated it to an unimportant section of a growing market. Its advantages were a long life, the ability to visually inspect the cell since the container was either of glass or (SAN) plastic (see Fig. 7.3), and historically its competitiveness against alternatives.

Drawbacks are the relative cost these days compared to newer cell designs [plates (see Fig. 7.4) had to be virtually handmade], its size and weight, and the need for a separate battery room with the incumbent problems of protection against acid and gas evolution.

Pasted plate cells are easier to manufacture. The plate is as shown in Fig. 7.5, the active material (in the form of a paste produced from lead oxide and dilute sulphuric acid) being pressed into the grid. These

COMPARISON DISCHARGE DATA BETWEEN : TUBULAR - PLANTE - VRLA CELLS

High Rate Discharge 340A to 1.69V/cell at 20 Deg.C.			Low Rate Discharge 53A to 1.8oV/cell at 20 Deg.C.			
Time (mins)	Tubular (V/cell)	Plante (V/cell)	VRLA (V/cell)	Time (hours)	Tubular (V/cell)	Plante (V/cell)	VRLA (V/cell)
1	1.795	1.840	1.945	Initial Volts	1.975	1.980	2.065
5	1.765	1.825	1.930	0.5	1.966	1.974	2.055
10	1.725	1.801	1.912	1.0	1.957	1.965	2.040
15	*	1.777	1.890	2.0	1.940	1.946	2.010
20	-	1.745	1.853	3.0	1.912	1.920	1.975
25	-	*	1.815	4.0	1.870	1.880	1.930
30	-	-	1.690	4.5	1.843	1.852	1.894
	* 14MIn - 1.69V.	*25.5Min - 1.69V		5.0	1.800	1.800	1.800

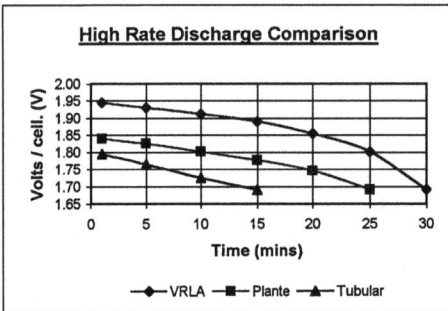

High Rate Discharge Comparison

—♦— VRLA —■— Plante —▲— Tubular

High Rate Discharge 340A to 1.69V/cell at 20 Deg.C.			
Time (mins)	Tubular (V/cell)	Plante (V/cell)	VRLA (V/cell)
1	1.795	1.840	1.945
5	1.765	1.825	1.930
10	1.725	1.801	1.912
15	1.690	1.777	1.890
20	-	1.745	1.853
25	-	1.690	1.800
30	-	-	1.690

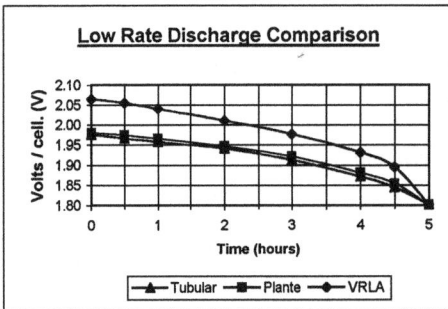

Low Rate Discharge Comparison

—▲— Tubular —■— Plante —♦— VRLA

Low Rate Discharge 53A to 1.8oV/cell at 20 Deg.C.			
Time (hours)	Tubular (V/cell)	Plante (V/cell)	VRLA (V/cell)
0	1.975	1.980	2.065
0.5	1.966	1.974	2.055
1.0	1.957	1.965	2.040
2.0	1.940	1.946	2.010
3.0	1.912	1.920	1.975
4.0	1.870	1.880	1.930
4.5	1.843	1.852	1.894
5.0	1.800	1.800	1.800

Figure 7.2 Comparison discharge data between tubular-Plante-VRLA cells.

cells are slightly smaller than Plante, and more competitively priced. They have a life of approximately 14 to 15 years against Plante life of 20 years plus. The cell still requires a separate battery room.

The tubular battery employs a rather different design of positive plate, this consists of what can best be described as the head of a rake (Figs. 7.6 and 7.7), the spines and the head of the rake being cast from a suitable lead-antimony alloy. The spines are surrounded with some form of semiporous plastic tube and the active material, lead oxide

YAP/YCP RANGE

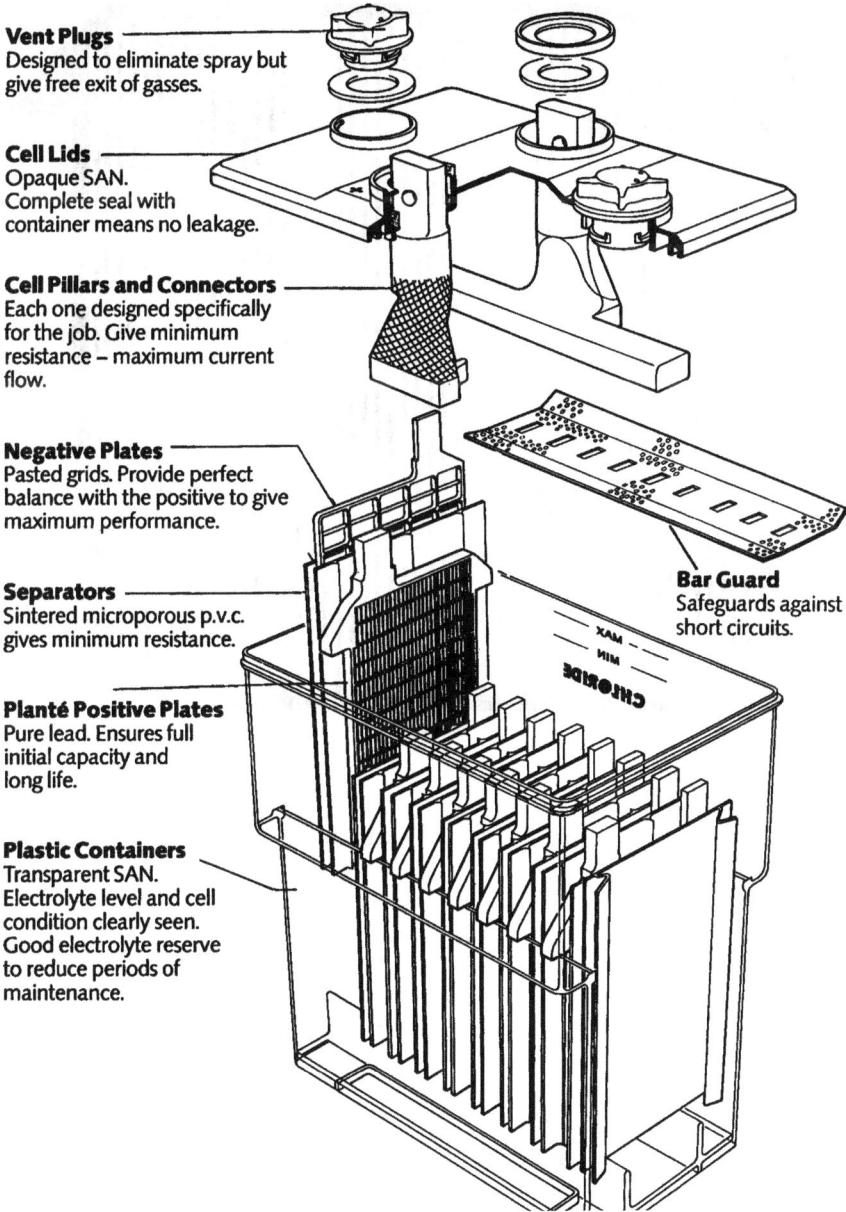

Vent Plugs
Designed to eliminate spray but give free exit of gasses.

Cell Lids
Opaque SAN.
Complete seal with container means no leakage.

Cell Pillars and Connectors
Each one designed specifically for the job. Give minimum resistance – maximum current flow.

Negative Plates
Pasted grids. Provide perfect balance with the positive to give maximum performance.

Separators
Sintered microporous p.v.c. gives minimum resistance.

Planté Positive Plates
Pure lead. Ensures full initial capacity and long life.

Plastic Containers
Transparent SAN.
Electrolyte level and cell condition clearly seen.
Good electrolyte reserve to reduce periods of maintenance.

Bar Guard
Safeguards against short circuits.

Figure 7.3   YAP/YCP range.

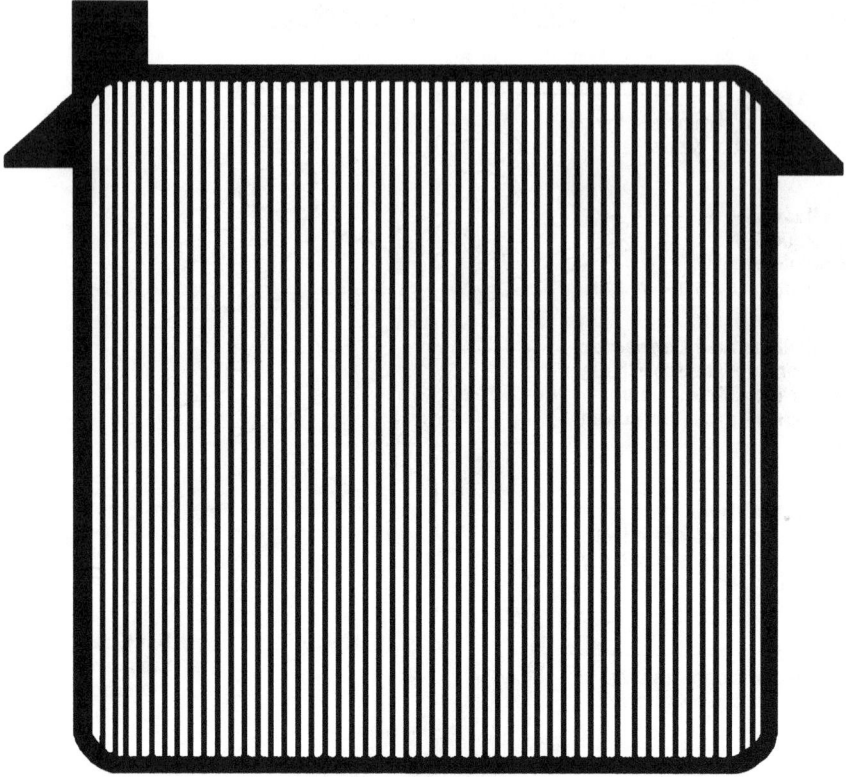

Figure 7.4

(PbO) in powder form, is vibrated down these cylinders. This is then converted to lead dioxide (PbO_2), the active material in the cell formation process. The resultant battery is expensive to produce and clearly has the same problems as the previous two designs. But it does have the advantage of being comparatively mechanically robust. The lead-antimony alloy gives added strength to the plate and this cell has been in use for traction purposes for many years. Its life depends on the design/application. Life is affected by the percent of antimony in the alloy as is the mechanical strength. The use of a low percentage anti-mony alloy is recommended for UPS applications, for example, 2 to 3 percent. This battery can find uses with UPS systems where physical conditions preclude the more popular VRLA cells. Such conditions include deep cycling discharges or large ambient temperature variations. These conditions can be encountered when designing a UPS system for telecommunication equipment that depends on solar and wind energy as a prime energy source.

Figure 7.5 Pasted plate.

All cells having flooded-type electrolyte construction clearly require a vent plug for each cell to allow release of gases, electrolyte replenishment, and access for measuring specific gravity and to inhibit acidic electrolyte spray. Basic designs are simple threaded plugs with a labyrinth construction inhibiting spray. There are designs, which in addition, enclose a small float indicating electrolyte level. Others may include a special flameproof feature and a small catalytic converter enabling exhaust gases to be reconverted to water, thus reducing the need to replenish electrolyte. Also, automatic refilling systems connected to a central electrolyte reservoir are available.

VRLA Types and Characteristics

The lead acid battery in most demand for utilization with UPS systems is undoubtedly the valve regulated type. The design was borne of a desire to provide an alternative battery for the telecommunications trade. Their desire was to break away from the very large central battery system and use smaller battery configurations for use with more numerous but smaller switching centers, The immediate design criteria

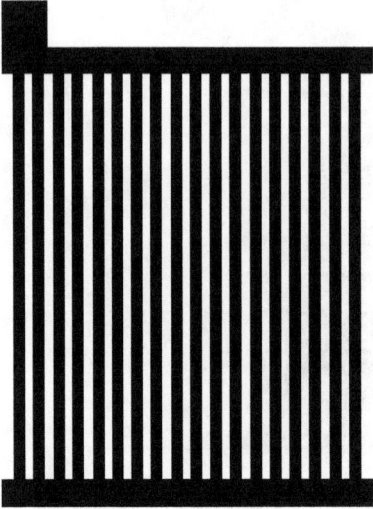

Tubular Positive Plate

Figure 7.6 Tubular positive plate.

Figure 7.7

was to reduce size, minimize gas evolution, and reduce water usage. The use of such cells for UPS applications came after this event, and the desire for very low maintenance was not really considered as a high priority.

The end result is a cell enclosed with a pressure valve, ensuring that there is a slight gas pressure. The issue of recombination of such evolved gases when the electrodes are surrounded with electrolyte was addressed by using a special very fine fiberglass mat [Note: This is often referred to as absorbent glass mat (AGM).] into which the electrolyte was soaked. The amount of electrolyte allowed in the cell is strictly limited, that is, no residual electrolyte can be seen or heard. During cell formation paths are created through the electrolyte held in the AGM to allow gas (oxygen) to migrate from the positive electrode to the negative electrode, where it recombines with hydrogen to form water.

The VRLA cell is attractive due to its size, some 50 percent smaller than Plante; it requires no special environment, that is, a battery room with room surfaces constructed to be protected against acid attack and also special ventilation to ensure gases evolved are evacuated; and its capital cost is some 50 percent less than Plante. In addition, since the cells are sealed there is little maintenance, that is, no checking of electrolyte levels is required. It does, however, have a shorter design life. This design life is usually 5, 10, or 15 years depending on construction and, of course, operating conditions including temperature and cycling. The cells are usually mounted vertically although there are some designs now available which can be operated horizontally, with front-mounted terminals, which can simplify stacking and, hence, save space.

VRLA clearly has the same chemical reactions as mentioned above and its construction is somewhat different than the flooded cells. A typical example is shown in Fig. 7.8. In fact the series of plates (see Fig. 7.9) are pasted with lead oxide (PbO)/dilute sulphuric acid which is then converted to lead dioxide as the cell is formed. As stated previously, there is no excess electrolyte held within the AGM (i.e., no residual liquid can be visually seen). For this reason the VRLA cell is often referred to as working on the starved electrolyte principle.The whole cell is sealed and a small valve is placed at the top of the cell to allow for any gas buildup to be expelled, the venting of gas under certain conditions occurring at above 1 atmosphere pressure. Note that this is not likely to occur except under fault conditions. Under normal operation the pressure in the cell is likely to be 3 lb per square inch (20684 Pa).

As previously mentioned there has been, for many years, a simple method to recombine the gases evolved in the chemical reaction, it involves the use of a catalyst to recombine the hydrogen and oxygen. Until recently this had only been employed on flooded-type cells, this

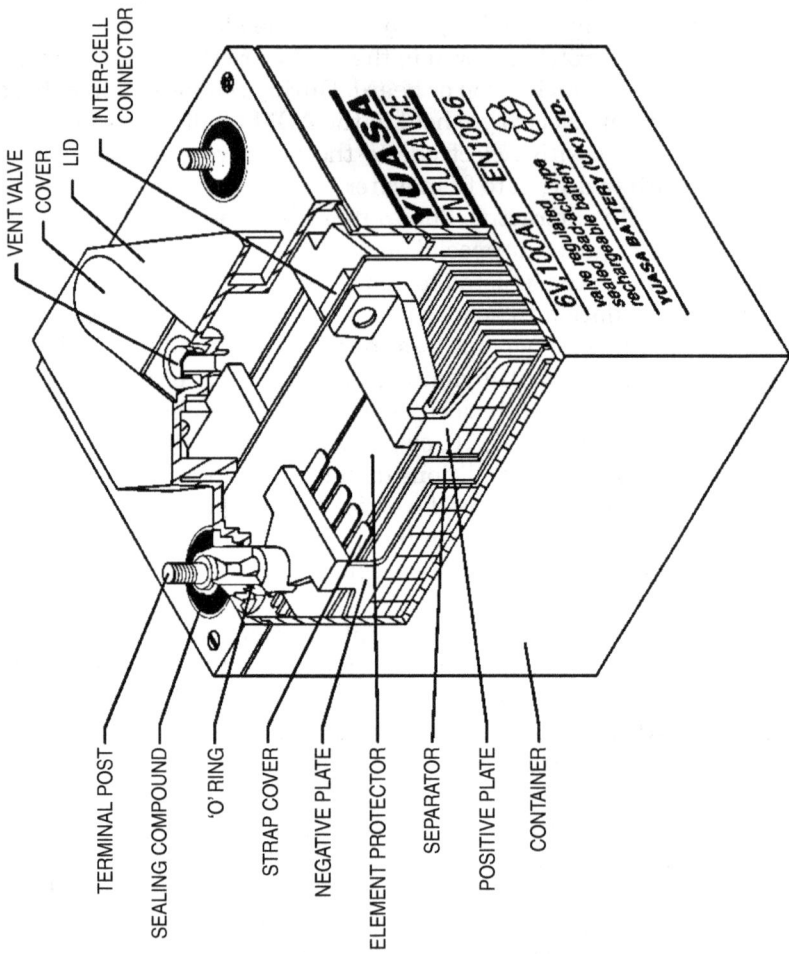

TERMINAL POST

SEALING COMPOUND

'O' RING

STRAP COVER

NEGATIVE PLATE

ELEMENT PROTECTOR

SEPARATOR

POSITIVE PLATE

CONTAINER

VENT VALVE

COVER

LID

INTER-CELL CONNECTOR

YUASA

ENDURANCE

EN100-6

6V 100Ah
valve regulated type
sealed lead-acid battery
rechargeable
YUASA BATTERY (UK) LTD.

Figure 7.8

194

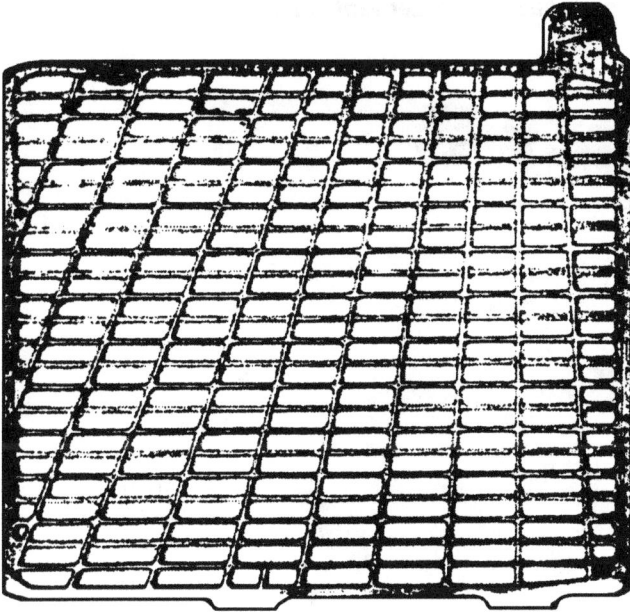

Figure 7.9

catalyst is now available in the form of a screw-in button to the top of VRLA cells as shown in Fig. 7.10. The suppliers claiming that electroyte levels are maintained and cell life is bolstered.

Low-pressure vent valves are designed to ensure that the pressure release is fairly tightly controlled and that the valve will reseal, and some of the valves are fitted with a flame-retardant system. See various types in Fig. 7.11.

In fact the life of the cell is basically governed by the amount of electrolyte present in the cell during manufacture. The losses are small, but present. As a rough estimate hydrogen loss is 10 ml per A/h per cell per year, under normal float conditions, a relatively small amount but attention is drawn to this as clearly a little ventilation is required. It should be noted that 4 percent admixture hydrogen to air is an explosive situation!

The charging of cells invariably leads to losses as before mentioned and it should be remembered that a slight but significant rise in local ambient temperature, say, $2°C$, is incurred in the charging cycle. It is recommended that cell blocks are allowed a 5- to 10-mm gap between blocks to allow air circulation.

The large majority of UPS designs use a constant-voltage charging system with current limit. The initial state of charge incurs a high current level. This gradually subsides to a low steady current when the

Liberty 2000 Maximizer vent Cut-away

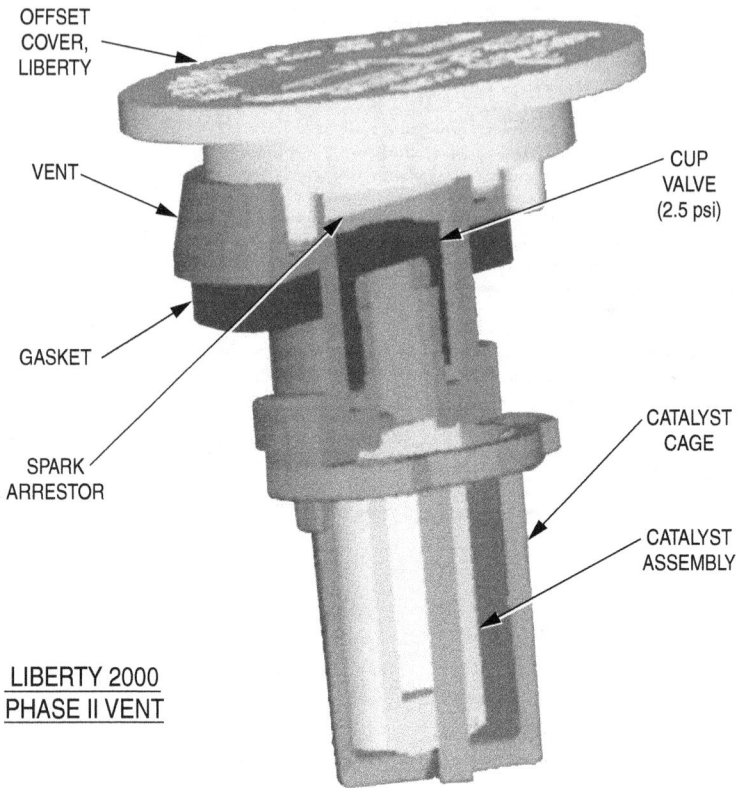

OFFSET
COVER,
LIBERTY

VENT

CUP
VALVE
(2.5 psi)

GASKET

CATALYST
CAGE

SPARK
ARRESTOR

CATALYST
ASSEMBLY

LIBERTY 2000
PHASE II VENT

Figure 7.10   Liberty 200 Maximizer vent cut-away.

float voltage is reached, usually 2.275 VPC at 20°C or 2.25 VPC at 25°C. (*Note*: There are slight variations here depending on the manufacturer's design). Figure 7.12 illustrates this (SOC = state of charge) and Fig. 7.13 also gives a good idea of variations to be expected with varying ambient temperatures.

An intermittent charging regime, claimed to be a benefit to VRLA life, consists of a cycling charging regime—2 to 3 days on float charge followed by 2 to 3 weeks idle, with the commitment to resume charging if battery voltage appears to decrease outside set tolerances. This system of intermittent charging claims to reduce the gradual decay in battery capacity and also increase the life of the cell. There is, however, conflicting evidence at this moment as to the efficacy of this system. Various other charging cycles are being tried, mostly varying the times

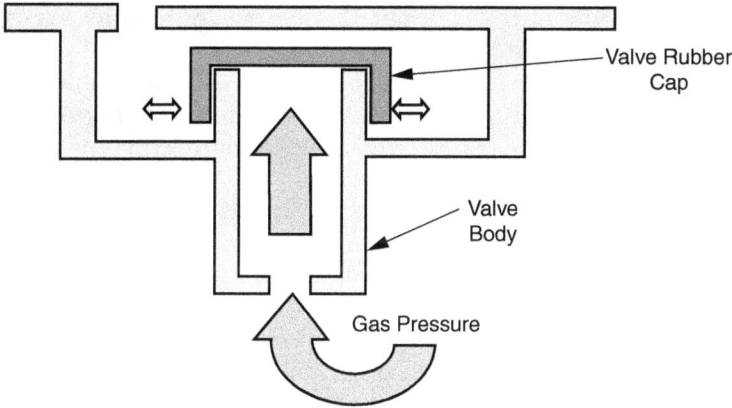

Figure 7.11   Vents.

Recharge Hours vs. Charging Voltage and Current

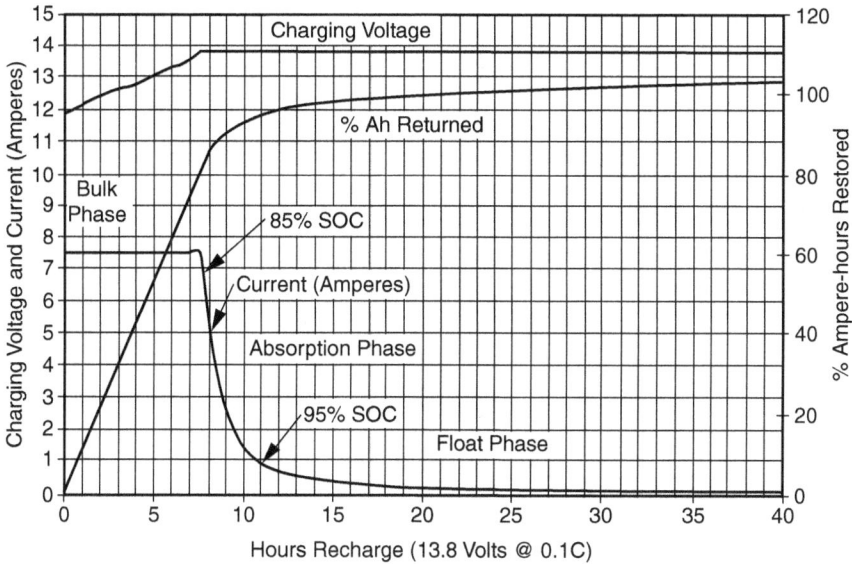

Figure 7.12   Recharge hours versus voltage and current.

in the cycling regime. Some experimental reports are in favor, others indicate an acceleration in hydrogen loss.

Recent developments have shown that short, high charging pulses have a restorative effect on lead acid performance. Such pulses destroy the build-up of sulphation on the plates. This ensures that the battery returns to near its original performance figures. This treatment may be repeated during the life of the cell, but this treatment should not be considered as a cycling event.

Ambient temperature has a significant effect on cell life. Published figures from Eurobatt give 10 years life at 20°C reducing to 5 years at an ambient temperature of 30°C. Figure 7.14 is for pure lead type cells, which claim a slightly better performance.

Ambient temperature also has a significant effect on cell performance manufacturers present figures based on either 20 or 25°C ambient temperature. Figure 7.15 is clearly based on 25°C. In Europe there is a tendency to base figures on 20°C.

Increasingly, UPS manufacturers are varying the charge voltage with ambient temperature, the basic formula being 0.003V/C per degree C. This temperature compensation device assists in attaining battery life, as illustrated in Fig. 7.16. For projects requiring cycling of cells such as solar or wind energy systems, there may be a need to use a higher constant charging rate.

Charging the cell requires a closely controlled input to achieve maximum results, and usually controlled rectifier systems are in common use. Clearly any rectification will produce a ripple in the dc output and this must be attenuated to 0.5 percent of dc float voltage and 0.5 C

CHARGING CHARACTERISTICS AT DIFFERENT TEMPERATURES

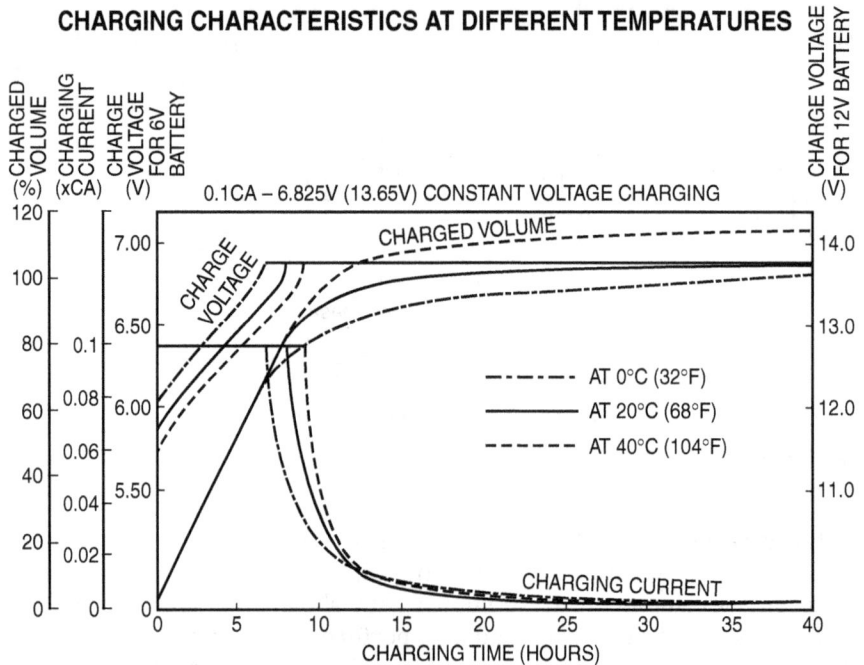

Figure 7.13 Charging characteristics at different temperatures.

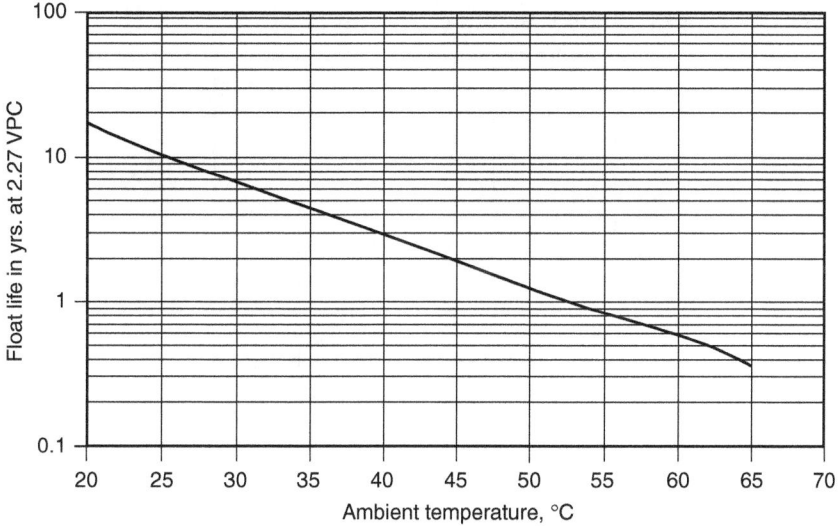

Figure 7.14

BATTERY LOAD CORRECTION FACTOR VS. BATTERY TEMPERATURE

BATTERY TEMPERATURE		BATTERY LOAD CORRECTION FACTOR				
C°	F°	15 min. rate	1 hour rate	5 hour rate	8 hour rate	20 hour rate
−9.4	15	.550	.580	.650	.705	.735
−6.7	20	.600	.630	.690	.735	.765
−3.9	25	.650	.680	.735	.765	.790
−1.1	30	.700	.725	.765	.790	.815
1.7	35	.740	.765	.800	.820	.840
4.4	40	.780	.800	.830	.845	.865
7.2	45	.820	.840	.855	.870	.890
10.0	50	.860	.865	.880	.895	.910
12.8	55	.875	.890	.910	.920	.930
15.8	60	.920	.930	.940	.945	.950
18.3	65	.940	.950	.955	.960	.965
21.1	70	.960	.970	.975	.975	.980
25.0	75	1.000	1.000	1.000	1.000	1.000
26.7	80	1.010	1.005	1.003	1.002	1.001
29.4	85	1.030	1.020	1.015	1.010	1.005
32.2	90	1.040	1.025	1.020	1.015	1.010
35.0	95	1.050	1.030	1.025	1.020	1.015
37.8	100	1.060	1.040	1.030	1.025	1.020

VRLA Battery Efficiency vs. Temperature and Discharge Rate

Figure 7.15   Battery load correction factor versus battery temperature.

Float Voltage Per Cell vs. Temperature

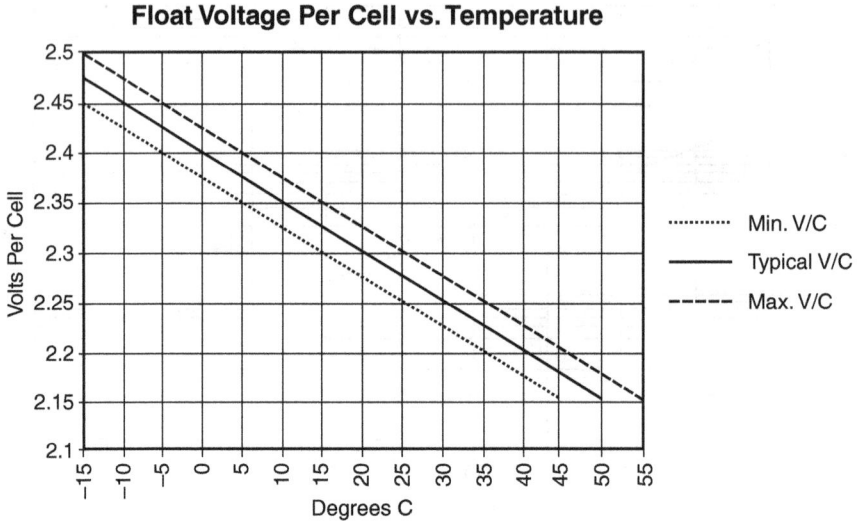

Figure 7.16   Float voltage per cell versus temperature.

amperes rms (5 amperes rms per 100 A/h of capacity) maximum. Ripple is also affected by frequency which maximizes at approximately 100 to 250 Hz. The effect of ripple is to cause heating of the cell, and the maximum overall heating effect should not be more than 5°C. The inverter clearly can also induce ripple onto the battery. However, the frequency of inverter switching is much higher and a simple choke is sufficient to solve this problem. Figures 7.17 and 7.18 illustrate, in the case of Fig. 7.17, load being correctly supplied by the charging system and in Fig. 7.18 some reversal of charging is occurring, *this is to be avoided*. Damage to cell plates may result. This problem will be the result of an incorrectly designed or adjusted charging system.

Most charging systems are designed to give current limit constant voltage. Thus, a cell on recharge initially will see a high current at a voltage considerably below the normal float voltage. As charging continues, voltage will rise to normal float volt level and the current will diminish to a low constant figure.

It should also be noted that the electrochemical efficiency of charging is high up to approximately 80 percent charge; thereafter the efficiency falls away and electrolysis occurs, resulting in hydrogen and oxygen evolving (see Fig. 7.19).

Life of VRLA cells is affected by the number of discharges (cycles) and the depth of such discharges. The cycling of a cell is also affected by the method of charging. For example, UPS systems normally float charge, and this is preferable, under the normal UPS operating conditions, to other methods. Temperature also has an effect, thus it is diffi-

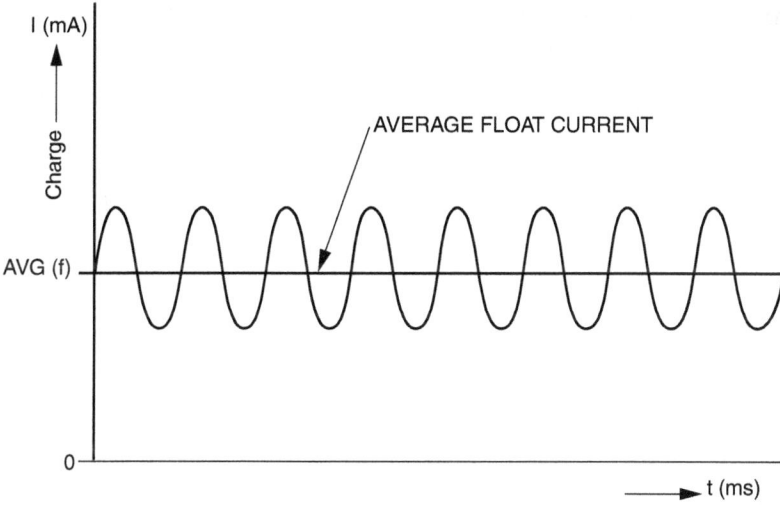

Float Charge with Low DC ripple current.

Figure 7.17   Float charge with low dc ripple current.

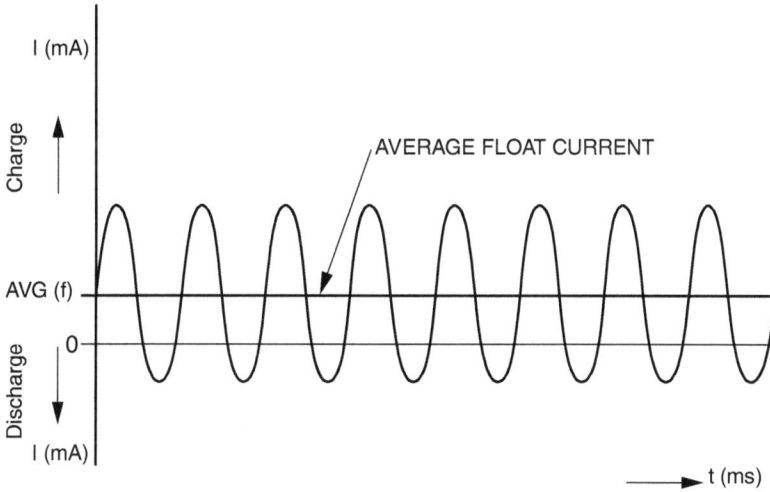

Float Charge with High DC ripple current.

Figure 7.18   Float charge with high dc ripple current.

cult to provide accurate figures. As a guide a typical 15-min autonomy at 20°C should provide at least 1000 cycles, this compares with a 1-h autonomy where, clearly, a deeper discharge occurs and the relevant figure would be 300 cycles. See Fig. 7.20.

Electrochemical efficiency related to battery state of charge

State of charge	Reactions	Electrochemical efficiency (% of current)
Maintenance in fully charged condition	(i) Discharge (self)	100% equivalent current
	(ii) Float recharge (compensation)	10%
	(iii) Electrolysis	90%
Approximately 80% state of charge	(i) Discharge	100%
	(ii) Recharge	100% at 80% SOC* — 10% at 100% SOC*
	(iii) Electrolysis	NIL at 80% SOC* — 90% at 100% SOC*
Fully discharged to approximately 80% state of charge	(i) Discharge	100%
	(ii) Recharge	100%

*SOC = 'state of charge'

Figure 7.19   Electrochemical efficiency related to battery state of charge.

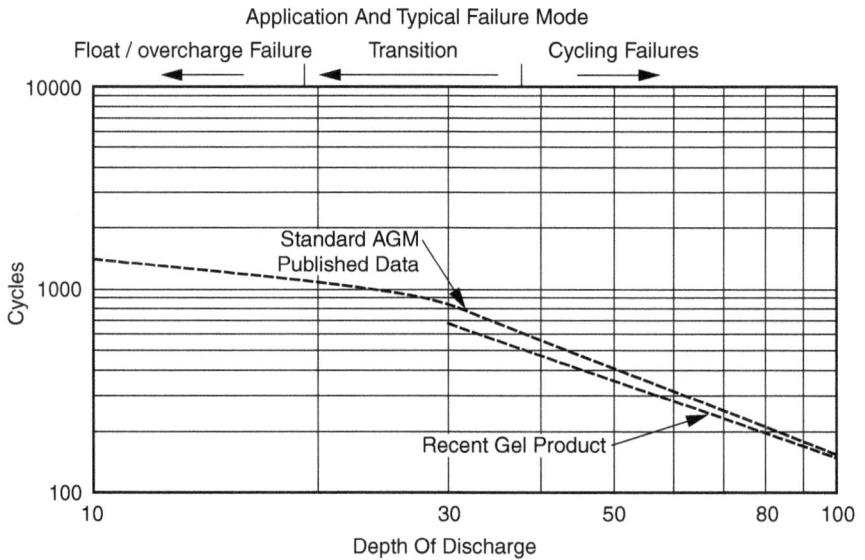

Figure 7.20   Application and typical failure mode.

Intermittent charging of VRLA cells is being tried with the aim of reducing the loss of hydrogen (thus, water loss). An example of this is to float-charge the battery for 2 to 3 days and then leave the battery in an open condition for 2 to 3 weeks and cycle in this mode with the proviso that an unacceptable voltage reading in the open cell condition will lead to an immediate recharge. Various cycling methods have been tried with varying success. So far as the writer can see, the jury is still out on this method.

With VRLA cells there is the danger of thermal runaway. This condition is the result of heat generated within the cell and results in accelerated dry out of electrolyte and also, in extreme cases, the melting or distortion of the plastic casing. The cause of this condition is inefficiencies in the cell due to I^2R losses and the exothermic oxygen recombination cycle. To avoid this problem consideration should be given to the following points:

1. High charging current

2. Unlimited or too high a charging current

3. Elevated float current

4. High ambient operating temperature

5. Unventilated battery enclosure

6. Badly designed battery enclosures with cells or blocks touching

7. Battery or system failures due to above conditions

8. Unrealistic life expectancy

Thermal runaway, now it is well understood, is not met as frequently as in past designs.

As a guide to minimize the temperature rise while minimizing recharge time, the figures below are desirable maximum figures.

Initial current limit as a function of 20-h rated capacity in ampere/h ©	Depth of discharge as a function of the 20-h rated capacity in ampere/h ©	Approx. discharge rate and period resulting in % depth of discharge
C/1	45%	15 min
C/2	55%	30 min
C/3	60%	3 h
C/5	90%	8 h

Where © is ambient temperature at 25°C.

Too high a charging voltage causes gassing since the amount of oxygen emitted exceeds the rate at which it can be diffused through the glass mat. Gassing rate is shown as a function of charge voltage in Fig. 7.21.

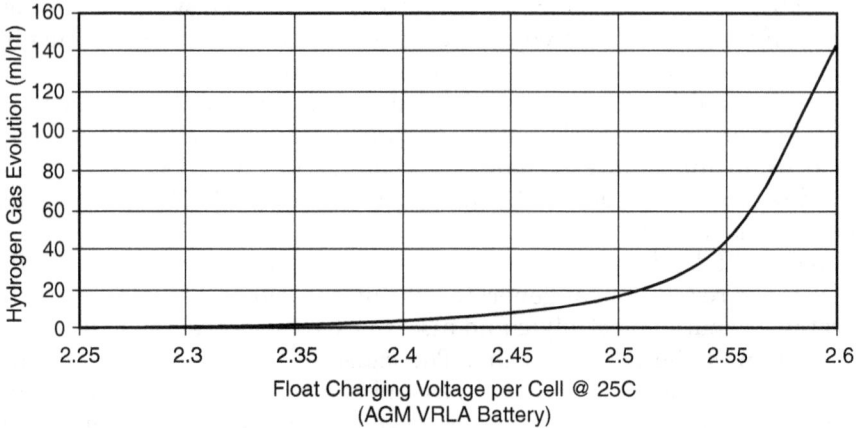

If the oxygen recombination cycle were 100% efficient there would never be any water loss from the electrolyte or gas emitted from the cell. However, the actual efficiency of the cells could vary from 90 to 99% under normal charging conditions so even though it is significantly reduced, there will be minor amounts of gas given off by the practical VRLA battery and some water loss from the electrolyte. However, with proper design and application, plate grid corrosion should be the eventual failure mode of the cell rather than water loss.

Figure 7.21

Most current-limited voltage-regulated charging systems will avoid such problems. However, it should be stated that grid corrosion and electrolyte dry out which occur with end of life conditions cause increased impedance/resistance in cells which can lead to this condition.

The plates are in most cases *not* pure lead. Small amounts of calcium, aluminum, and silver aid manufacturing (lead castings are subject to damage due to their malleability) and also aid charging regime and inhibit corrosion, and in the case of silver assist in ensuring a high discharge rate. When producing these alloys, the crystal structure of the material is quite important. Alloys with large crystals are to be avoided, as this assists the electrolyte in developing corrosion, the spaces between large crystals allowing the electrolyte to increase corrosion activity. Historically, antimony was used to strengthen the plate and indeed is still used mostly in the tubular cell construction for traction applications. However, there is a tendency during the life of the cell for the antimony to leach out of the plate and form a masking surface on the negative plate, thus reducing the cell performance. The traction applications usually call for deep discharges and a relatively short life and antimony is still used for such projects. However, when tubular cells are used for UPS activities where deep discharges are required but there is no problem of mechanical shock the antimony is reduced from 7 or 8 percent to 2 percent.

Calcium is added to the plate in small quantities, about 0.8 percent for the same reason, that is, to strengthen the plate during its manufacturing process. Higher percentages of calcium have undesirable effects, such as corrosion in the form of grid growth. And in many cases this has itself been superseded by an alloy incorporating the use of tin, again in small quantities (0.6 percent). This alloy increases the grain size of the metal, the larger the grain size the lower the corrosion at the positive plate.

Very small amounts of silver added to the alloy have the effect of reducing resistance thus increasing the cell's ability to give a high rate discharge. Clearly, such amounts are very small due to cost.

The charging of cells has, until recently, been based on a float charge system, usually based on 2.27 V per cell (VPC) with current limit. By and large, UPS systems employ constant voltage with current limit circuits. But it has now been proved that short periods of charge followed by longer periods of open circuit with no charge are beneficial to cell life. The advantages are a reduction in corrosion of the plates, which occurs during charging, and loss of electrolyte. Claims of 30 to 50 percent increase in life are made.

Plastic containers vary in material used depending on application (Fig. 7.22), ABS (acrylonitrile-butadiene-styrene) being the favored material to meet the various flameproof standards such as BS 6290 and UL94V-0. There are, however, other flameproof materials available such as some grades of polypropylene which are approaching the same flameproof standards but at a lower cost.

In the construction of the battery block, it should be noted that most applications require the battery to provide power for a relatively short time (e.g., 15-min autonomy period). Clearly, the plates and, indeed, the collection of power from the plates through the post thus have to provide a low resistance path. In practice the post, on a full power discharge, can generate heat and in some circumstances it may be preferable to fit a brass insert within the post to reduce resistivity, to provide a robust connection point, and to lower temperature. See Fig. 7.23a.

In addition to the above AGM variety of VRLA, a variation employing pure lead plates is available. This cell operates at a slightly higher pressure, and since pure lead is employed corrosion is reduced, charging rate can be dramatically improved, and its temperature tolerance is improved (e.g., at 30°C design life is approximately 7 years). The cycling, that is, the number of charging/discharging events, is also improved, especially for the periodic, small discharges incurred on most UPS applications.

In addition to the VRLA cells described above there is the gel-type VRLA cell which differs in construction. The glass mat is replaced by a fumed silica thickening agent in the electrolyte. The mix, in liquid form, is poured into the cell. This substance then gels. During manu-

Characteristic	Hard Rubber	SAN	Polycar-bonate	PVC	ABS	Polypro-Plyene
Durability		Excellent	Excellent	Excellent	Excellent	Excellent
Rigidity		Excellent	Excellent		Excellent	
Impact Resistance			Excellent	Excellent		
Typical Case to Cover Bond		Epoxy	Epoxy	Epoxy	Epoxy	Thermal Weld
Typical Terminal Seal						
Transparent		Yes	Yes	Yes	No	No
Flame Retarding		No	Yes	Yes	Optional	Optional
Gasses Evolved when Burned						
Vented Cells		X	X	X		X
VRLA Cells					X	X
Typical Applications		20 yr. Stat.	20 yr. Stat.	20 yr. Stat.	20 yr. Stat. 5-10 Yr. Stat.	SLI Traction 5-10 Yr. Stat.

Figure 7.22

facture the finishing charge evolves excess water, and due to this process fissures are established in the gel between the plates. It is through these fissures that the oxygen percolates during charging, traveling from the positive to negative plate. Since silica is added to the electrolyte it is clear there is slightly less acid available, thus lower capacity. However, the gelled electrolyte has nearly 100 percent contact with the plates and the walls of the cell, thus good thermal conductivity is a result. Thermal run-away susceptibility is reduced and such cells are better suited to high ambient temperature areas. The addition of phosphoric acid to the electrolyte will double the cycle life capabilities but give a reduced float service life.

Development of the VRLA type of cell continues. For example, new methods are being used to create the plate: Instead of being cast, manufacturers are attempting to form plates from a strip of lead, punching out sections, thus providing a continuous strip which may be cut to the required length. The result appears similar to Fig. 7.9. The resultant plate is then stretched slightly and the lead-oxide paste mixture added. The aim is to produce a 5-min autonomy period cell, and this probably will be created by making plates thinner and increasing the porosity of the active plate material. Porosity increase will allow a better availability of the material to the electrolyte. There is also a demand for cells with increased life, thus reducing maintenance costs. The aim here is for a cell with a good 10 year+ life at 27 to 29°C (80 to 83°F).

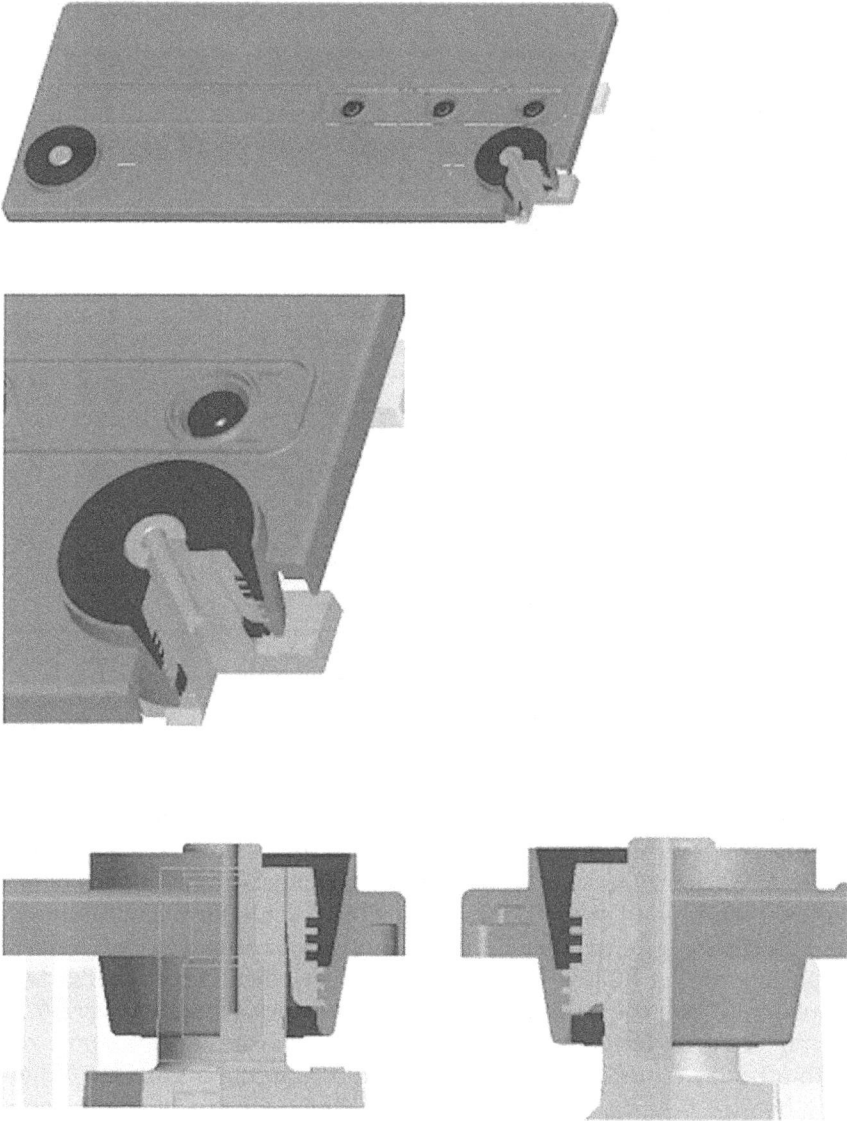

Figure 7.23a

Battery Monitors

The VRLA cells are predominantly chosen for many UPS applications due to their advantages in size, low initial cost, low maintenance, and the attraction of being able to forego the battery room environment.

However, the VRLA cells have a relatively short life, are temperature sensitive, and since cells are sealed it is difficult to assess the health of

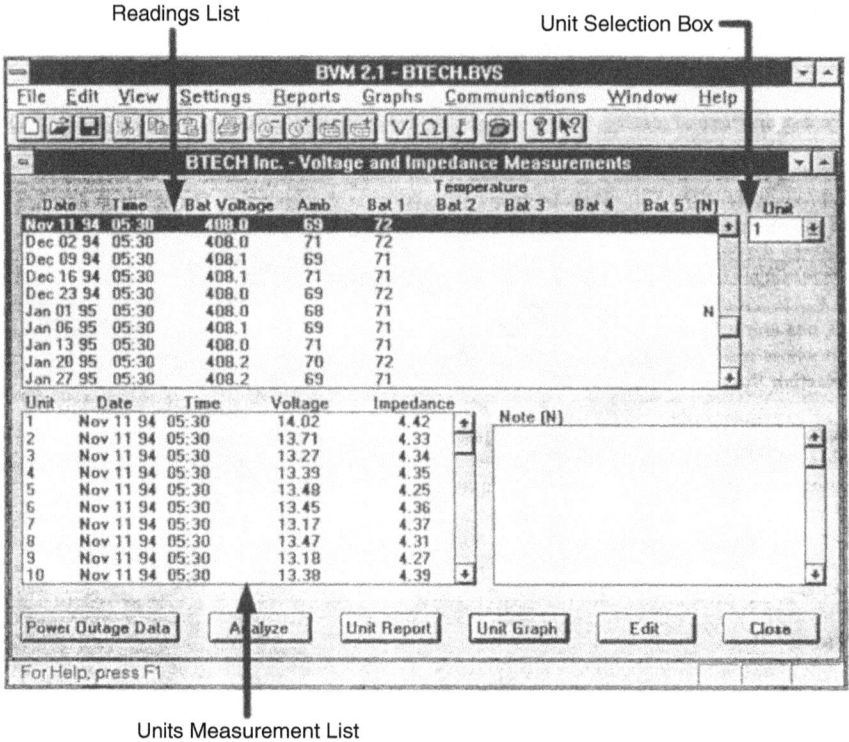

Readings List

Unit Selection Box

BVM 2.1 - BTECH.BVS

File Edit View Settings Reports Graphs Communications Window Help

BTECH Inc. - Voltage and Impedance Measurements

Date	Time	Bat Voltage	Amb	Temperature Bat 1	Bat 2	Bat 3	Bat 4	Bat 5 [N]		Unit
Nov 11 94	05:30	408.0	69	72						1
Dec 02 94	05:30	408.0	71	72						
Dec 09 94	05:30	408.1	69	71						
Dec 16 94	05:30	408.1	71	71						
Dec 23 94	05:30	408.0	69	72						
Jan 01 95	05:30	408.0	68	71						
Jan 06 95	05:30	408.1	69	71				N		
Jan 13 95	05:30	408.0	71	71						
Jan 20 95	05:30	408.2	70	72						
Jan 27 95	05:30	408.2	69	71						

Unit	Date	Time	Voltage	Impedance	Note [N]
1	Nov 11 94	05:30	14.02	4.42	
2	Nov 11 94	05:30	13.71	4.33	
3	Nov 11 94	05:30	13.27	4.34	
4	Nov 11 94	05:30	13.39	4.35	
5	Nov 11 94	05:30	13.48	4.25	
6	Nov 11 94	05:30	13.45	4.36	
7	Nov 11 94	05:30	13.17	4.37	
8	Nov 11 94	05:30	13.47	4.31	
9	Nov 11 94	05:30	13.18	4.27	
10	Nov 11 94	05:30	13.38	4.39	

Power Outage Data   Analyze   Unit Report   Unit Graph   Edit   Close

For Help, press F1

Units Measurement List

Figure 7.23b

the cells. Gone are the days when measurement of specific gravity of the electrolyte and, for some types, a visual indication were available.

Thus, the user has to assess battery characteristics by means of circuitry. Virtually all static UPS systems have some form of simple battery monitoring for smaller ratings simply indicating a battery failure or an impending failure. Larger sets will periodically test the battery by phasing back the rectifier, allowing the battery, for a short period, to supply the load. Over this short period the discharge curve is compared with a known standard similar curve, thus gauging a fair idea of battery health. Clearly, a refined testing system would add to the confidence in the battery. Such tests do not give conclusive proof that the battery will perform for the designed autonomy period, and to utilize the battery for such tests may impede its efficiency to support the system if a mains failure should occur soon after the test!

Measurements of voltage and current, and now impedance, are clearly of assistance, and thus a market has grown for a sophisticated battery monitor.

Clearly, a monitoring system providing a continuous output signal entailing all the various properties of the cells is essential. Such devices are now available but at some cost, and there has now developed two schools of thought:

1. Use the existing simple built-in battery monitoring system in the UPS and replace the cells say every 5 to 6 years for a VRLA battery with a design life of 10 years.

2. Install a complete monitoring system, maintain the system replacing cells as and, when required, obtain a longer battery life.

The disadvantage of the former is that rogue cells may not be noticed, and such cells will induce weaknesses in their immediate neighbors. The advantage of course is cost!

Thus, a range of battery monitors is now available which measure continuously individual cell/blocks for voltage, impedance, current, and temperature variations. Such systems are sophisticated and require consideration. It should be noted that IEC is preparing a specification for such devices under the title IEC 62060.

Monitors usually take the form of a wall-mounted receiving station to which are connected small signal wires from each cell block. These can take the form of a fuse-protected wire or a glass-fiber cable. There is an alternative available which runs the test wire in series through the entire battery, thus eliminating multicabling. The system measures current, voltage, impedance, temperature, and variations during normal float and discharge and compares such results with known results for a healthy battery irrespective of battery age. The figures are computed to form a trend of cell performance enabling action to be taken to forestall any failures. The system may also be interrogated by a normal office PC or via modem telephone link to a central maintenance area. (See Figs. 7.23 to 7.27.)

Nickel-Cadmium Cells

Chemical Reactions

Nickel-cadmium batteries work under the chemical equation as shown below:

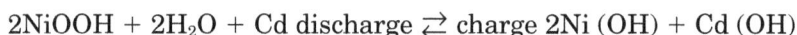

$$2NiOOH + 2H_2O + Cd \text{ discharge} \rightleftarrows \text{charge } 2Ni(OH) + Cd(OH)$$

The positive plate uses nickel hydroxide as the active material, and the negative plate uses cadmium hydroxide.

The cell utilizes as an electrolyte a solution of potassium hydroxide with a small amount of lithium hydroxide to improve cyclic and high-temperature performance. The solution is used purely as a method of ion transfer and no chemical change to the solution occurs. The support structure for the active material is steel which is unaffected by the chemical activity within the cell.

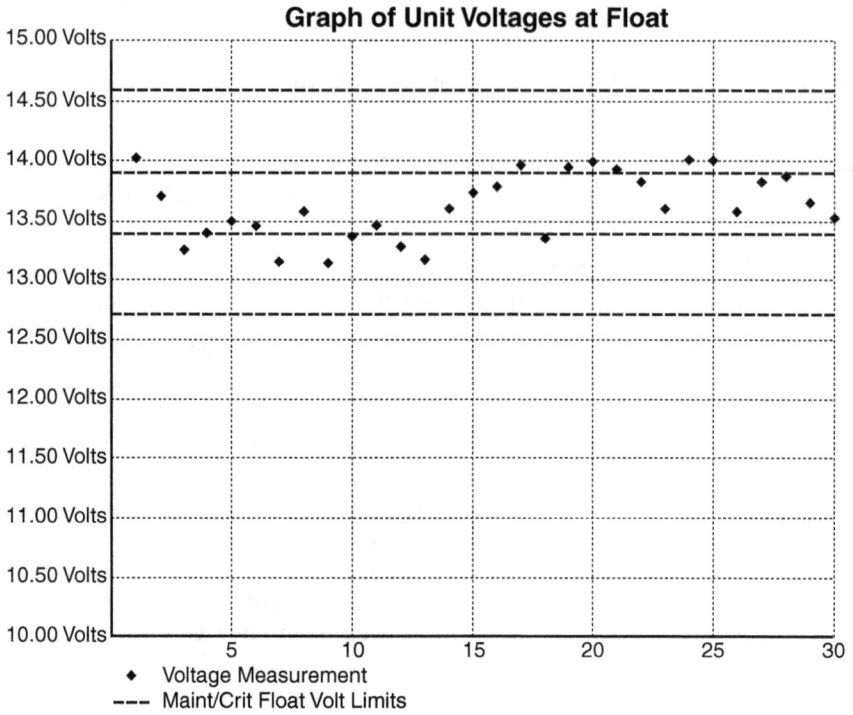

Figure 7.24 Graph of unit voltages at float.

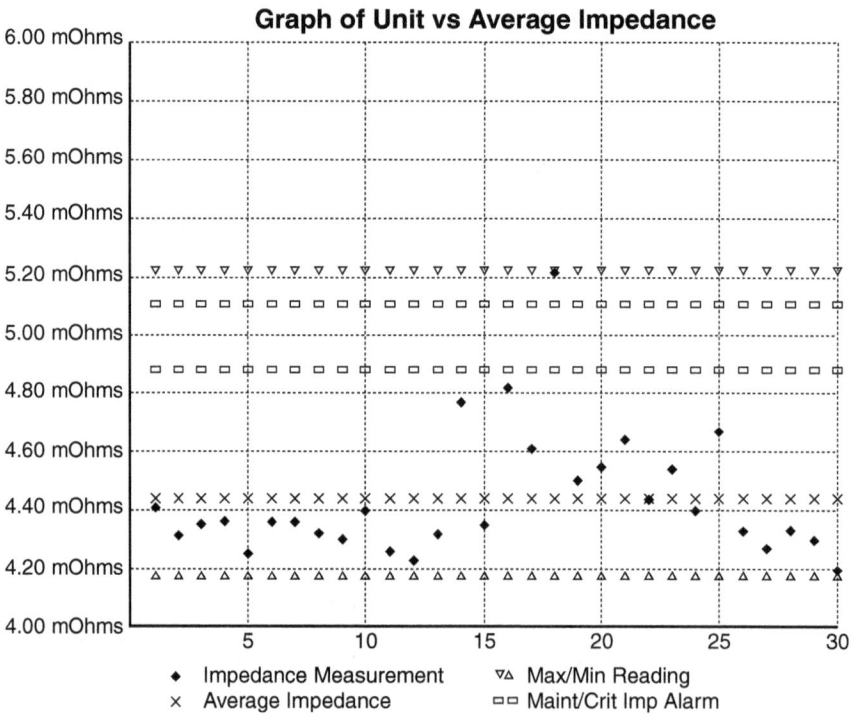

Figure 7.25 Graph of unit versus average impedence.

Trend Graph of Unit 18 Voltage

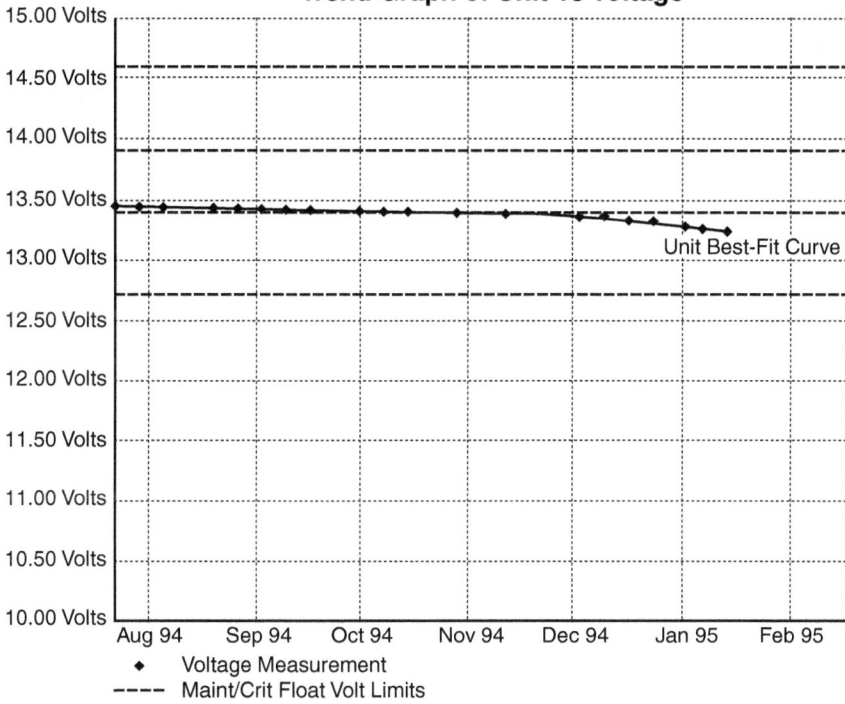

15.00 Volts	
14.50 Volts	
14.00 Volts	
13.50 Volts	
13.00 Volts	Unit Best-Fit Curve
12.50 Volts	
12.00 Volts	
11.50 Volts	
11.00 Volts	
10.50 Volts	
10.00 Volts	

Aug 94   Sep 94   Oct 94   Nov 94   Dec 94   Jan 95   Feb 95

♦   Voltage Measurement
----   Maint/Crit Float Volt Limits

Figure 7.26   Trend graph of unit 18 voltage.

Trend Graph of Unit 18 Impedance (Average)

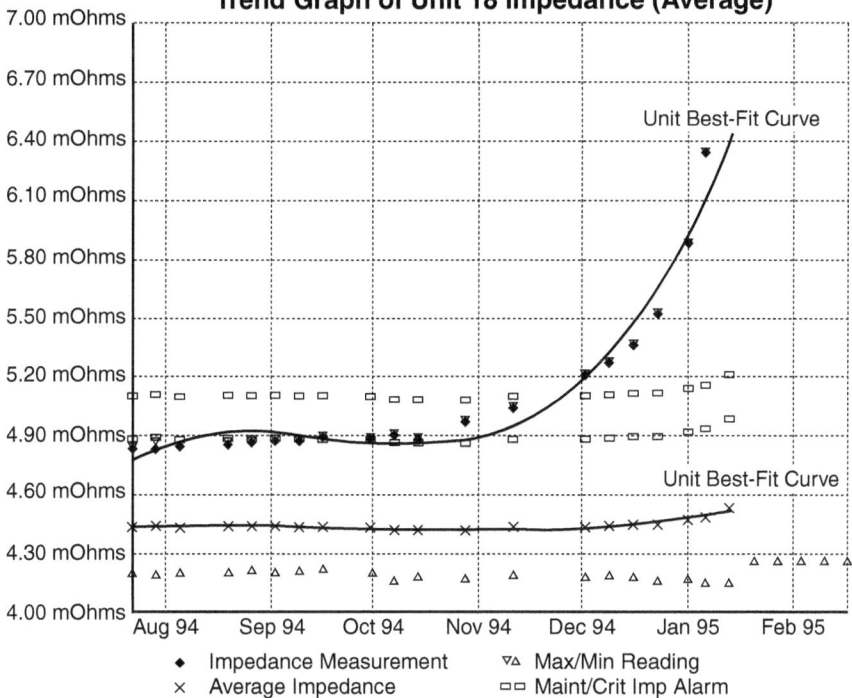

7.00 mOhms	
6.70 mOhms	
6.40 mOhms	Unit Best-Fit Curve
6.10 mOhms	
5.80 mOhms	
5.50 mOhms	
5.20 mOhms	
4.90 mOhms	
4.60 mOhms	Unit Best-Fit Curve
4.30 mOhms	
4.00 mOhms	

Aug 94   Sep 94   Oct 94   Nov 94   Dec 94   Jan 95   Feb 95

♦   Impedance Measurement   ▽△ Max/Min Reading
×   Average Impedance   ▫▫ Maint/Crit Imp Alarm

Figure 7.27   Trend graph of unit 18 impedance (average).

Pocket Plate

There are two basic types of plate structure, a thick plate comprising perforated steel sheets between which the active material is compressed, usually in the form of pockets or tubes. This method of construction affords cells with a relatively long discharge period (see Fig. 7.28*a* and *b*).

Construction
Nickel-cadmium pocket plate

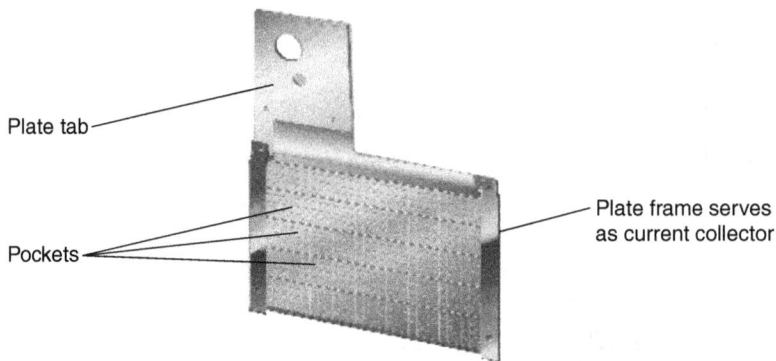

Plate tab

Pockets

Plate frame serves as current collector

Final pocket plate assembly

Figure 7.28   (*a*) Construction of nickel-cadmium pocket plate.

Construction

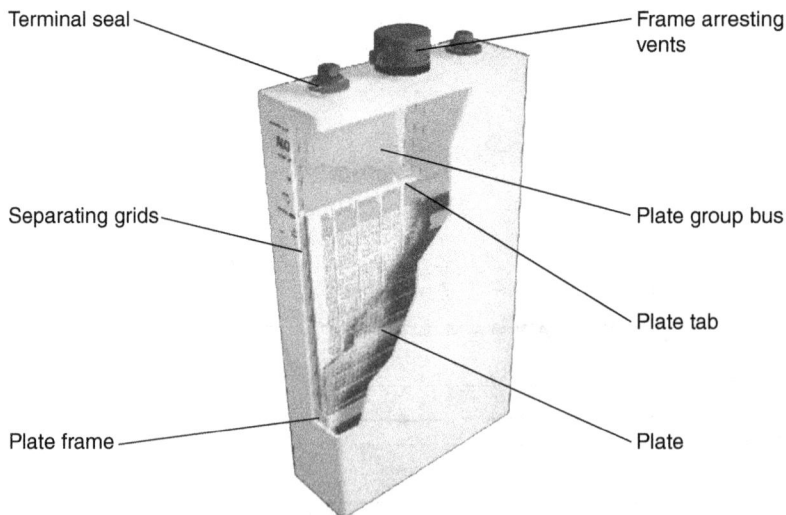

Terminal seal

Separating grids

Plate frame

Frame arresting vents

Plate group bus

Plate tab

Plate

Figure 7.28   (*b*) Construction.

Sintered Plate

The alternative method of construction uses sintered metal plates. These are constructed using nickel-plated steel onto which nickel powder is sintered and then nickel hydroxide is impregnated into the porous nickel structure. The negative plate (cadmium) is constructed by mixing together the active material, plastic bonding material, and spreading this onto a perforated nickel-plated steel strip. Such designs give a high discharge rate. See Fig. 7.29.

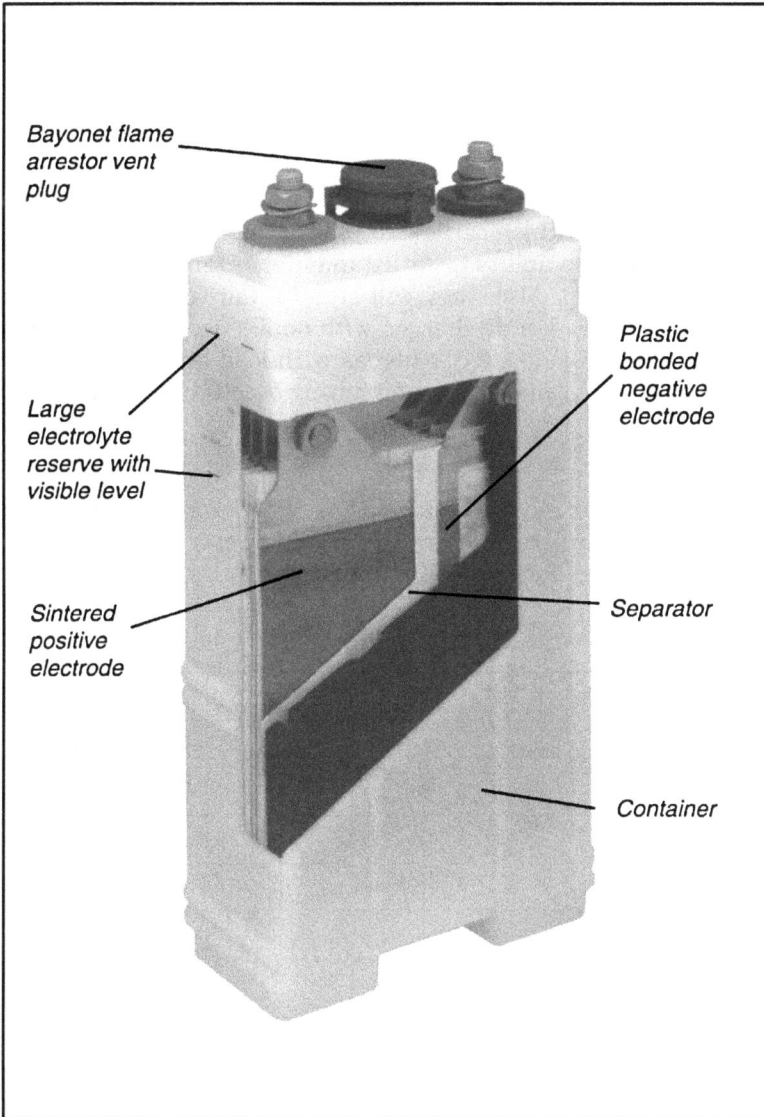

Figure 7.29

Since the electrolyte is acting only as an ion exchange medium, it is possible to increase the reservoir of electrolyte and thus reduce maintenance.

The plate separators consist of a microporous polymer/nonwoven felt sandwich, designed to ensure complete plate separation and good design life.

Sealed Type

Nickel-cadmium cells are available in both open-type and sealed versions. Their advantages are long life and ability to withstand repeated cycling and also to work with a wider ambient temperature range (-40 to $+60°C$). See Fig. 7.30. The main disadvantages are its high initial cost and, compared to VRLA cells, size, and for the open-type nickel-cadmium, the need for a well-designed battery room. Advantages for the open type are long life, which can be 20 years at normal ambient temperatures ($25°C$), and its ability to withstand higher temperatures than lead acid (see Fig. 7.31). Also, they will stand a fair amount of rough treatment: Cells may be left discharged with no ill effect. Overcharging or undercharging causes no effect whereas with lead acid cells there is the danger of plates deforming. Short-circuiting of cells can be tolerated.

Charging of Cells

Charging of cells under most UPS conditions will occur at 1.41 to 1.45 VPC using a voltage-controlled current-limited circuit (see Fig. 7.32). End of discharge occurs for most UPS applications at about 1.06 VPC.

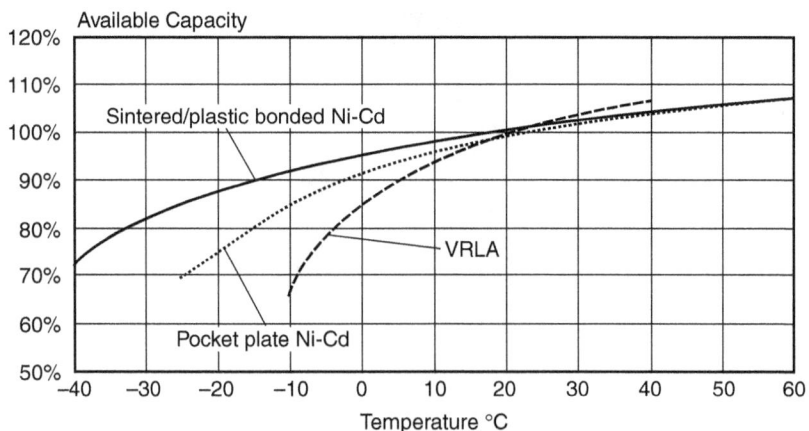

Effect of temperature on performance

Figure 7.30 Effect of temperature on performance.

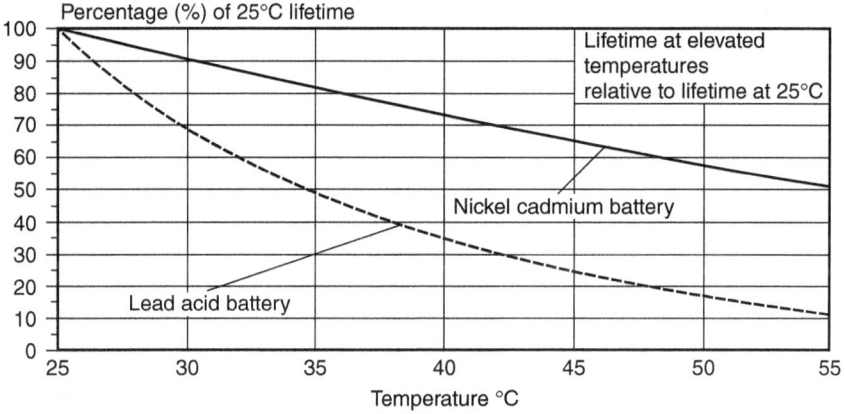

Figure 7.31 Lifetime at elevated temperatures relative to lifetime 25°C.

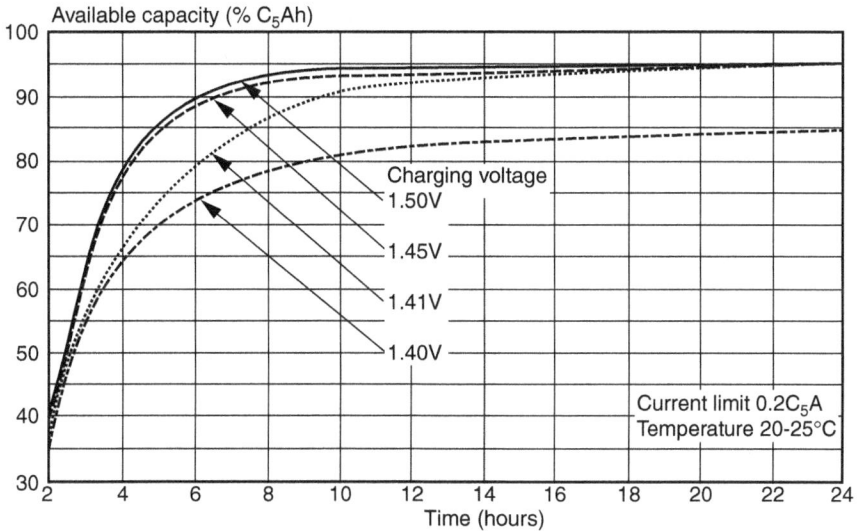

Capacity available for typical charging voltages

Figure 7.32 Capacity available for typical charging voltages.

Charging cells is quite efficient until approximately 80 percent charge has been accepted. Thereafter, charging efficiency falls off and, increasingly, gas evolution takes place. See Fig. 7.33.

Ripple apparent from either the charging or inverter circuits has virtually no effect on the nickel-cadmium cell. There may be a very slight increase in water dissipation but for a normal UPS circuit this problem is not likely to be met.

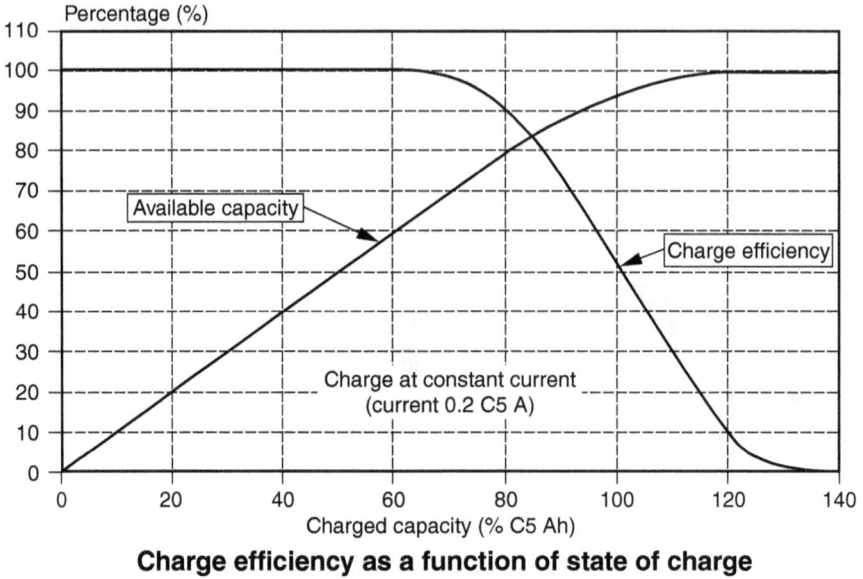

Charge efficiency as a function of state of charge

Figure 7.33 Charge efficiency as a function of state of charge.

Evolution of gas (both hydrogen and oxygen) is the result of the charging mechanism. At the float voltages normally employed, it is not too great. Take as an example: 280 cells of 70A/h placed on a two-step three-tier stand in a room 2m × 10m × 3m. Assume charging at 0.25, thus charging current is 56 amperes. Now, 1A/h will break down 0.366 cm^3 of water and since 1 cm^3 of water will produce 1.865 liters of gas, of which two-thirds are hydrogen, then 1 A/h will produce 0.42 liters of hydrogen. Thus, the volume of hydrogen evolved from the said 70 A/h battery system is:

280 (no. of cells) × 56 (charging current) × 0.00042 m^3 (of hydrogen)
= 6.59 m^3 of hydrogen per hour.

The total volume of room is 2 × 10 × 3 = 60 m^3. Since the stand takes up space in the room, in this case 11 m^3, the total volume of free air in the room is 49 m^3. Thus, the total volume of hydrogen evolved after 1 h charging at 0.25 expressed as a percentage is 6.59/49 = 13.5 percent.

To keep the concentration of hydrogen down to 3%, the number of air changes in the room shall be 13.5/3 = 4.5 changes of air per hour. It should be noted that a 4% concentration of hydrogen is an explosive mixture.

Discharging and charging of cells, that is, cycling is certainly superior to lead acid cells. Figure 7.34 illustrates the performance of pocket

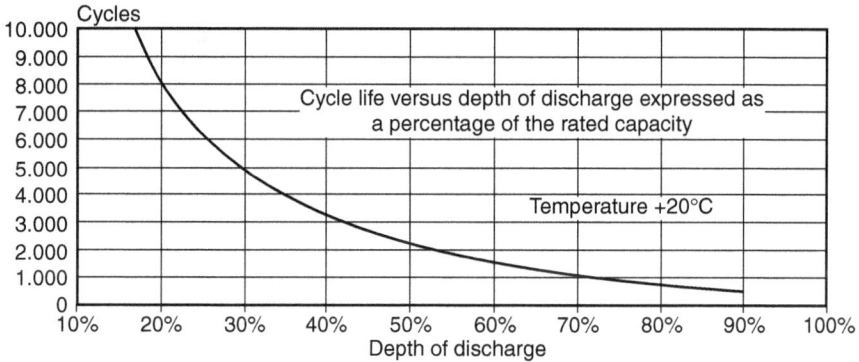

Typical cycle life versus depth of discharge for pocket plate battery

Figure 7.34 Typical cycle life versus depth of discharge for pocket plate battery.

plate cells, for the sintered plate design we can expect 3500 cycles at a depth of discharge of 80 percent.

Thermal runaway is not a problem with nickel-cadmium cells, the majority of the charging energy is stored in the battery and the chemical reaction taking place is slightly endothermic (cooling action). However, as gassing starts to occur at about 80 percent of full charge, the charging efficiency falls away and the energy is used to decompose water and also evolve some heat. But, as can be seen from Fig. 7.35, such heating is likely to be very low due to the charging voltage normally used on UPS systems, 1.4 to 1.45 VPC.

Conversely, heat evolved during discharge may be calculated by using the formula in Fig. 7.36.

Operating nickel-cadmium cells at high ambient temperatures, say above 40°C, may lead to the potassiun hydroxide carbonating, that is, forming crystals of potassium carbonate and reducing the cell efficiency. The only solution is to refill completely every cell with new electrolyte. Note this does not occur with sintered plate cells.

Electrolyte topping up should occur only every 10 years and longer still, say every 15 years, with the *sealed cells*. It is inadvisable to operate the sealed nickel-cadmium cells above 40°C.

The sealed nickel-cadmium cell is still a wet cell and its advantages are very long life, significantly reduced gas production, and even lower maintenance than the other nickel-cadmium cells. See Fig. 7.37.

The cell uses a pocket-plate design but plates are designed to reduce water decomposition. Normally, the charging process incurs oxygen evolution at the positive plate and hydrogen evolution at the negative plate. Oxygen commences to be produced just prior to the fully charged state occurring, and at fully charged state the result of charging is merely the production of oxygen.

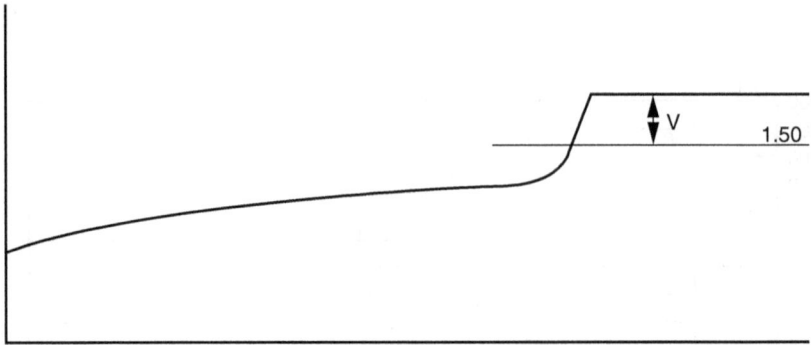

If for example the charging is done with a maximum constant voltage of 1.70 VPC, use 1.70 − 1.50 = 0.2 V in the following formula:

$W = V \times I \times n$ where

W = Dissipated power in watt

I = Charging current in Amps

n = number of cells in series connection

V = Difference between charging voltage per cell and 1.50 V

The charging current is dependent upon how high the final voltage is allowed to go, and on the cell type.

Figure 7.35

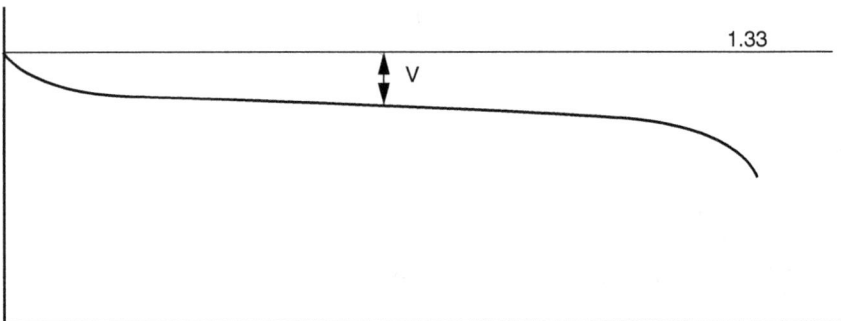

The discharge voltage is of course different for different rates of discharge and different cell types. Use the following formula:

$W = V \times I \times n$ where

W = Dissipated power in watt

I = Charging current in Amps

V = Difference between 1.33 V and the discharge voltage per cell

Figure 7.36

Vantage   **Construction features**

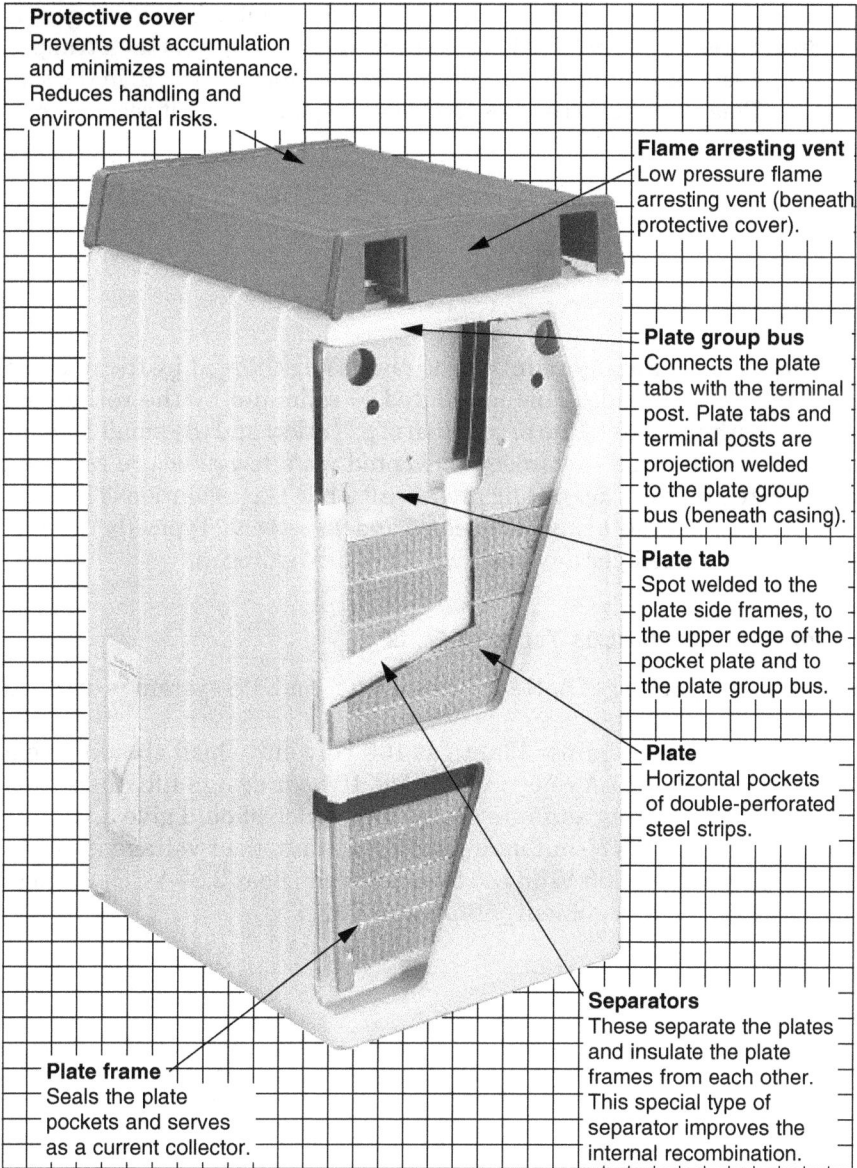

Protective cover
Prevents dust accumulation
and minimizes maintenance.
Reduces handling and
environmental risks.

Flame arresting vent
Low pressure flame
arresting vent (beneath
protective cover).

Plate group bus
Connects the plate
tabs with the terminal
post. Plate tabs and
terminal posts are
projection welded
to the plate group
bus (beneath casing).

Plate tab
Spot welded to the
plate side frames, to
the upper edge of the
pocket plate and to
the plate group bus.

Plate
Horizontal pockets
of double-perforated
steel strips.

Separators
These separate the plates
and insulate the plate
frames from each other.
This special type of
separator improves the
internal recombination.

Plate frame
Seals the plate
pockets and serves
as a current collector.

Figure 7.37 Vantage™ construction features.

The negative plate is better able to accept charge and hydrogen production only occurs when the plate is fully charged. So the cell is constructed with a larger than normal negative plate. Thus, oxygen is produced prior to the production of hydrogen. This oxygen is then trapped in porous separators and eventually displaces some electrolyte in the separators. This eventually reaches the negative plate where it reacts chemically as follows:

$$2Cd + O_2 + 2H_2O = 2Cd\,(OH)_2$$

or electrochemically

$$O_2 + 2H_2O + 4E = 4OH$$

Thus, some of the cadmium on the plate is discharged by the reaction to cadmium hydroxide. This is reduced to cadmium by the recharging activity within the cell. The production of gas is low and the small amount of gas evolved (mostly hydrogen) is vented via a low-pressure reclosing flame-arresting vent, operating pressure 0.2 bar. Gas evolution is limited, recombination levels being between 85 and 95 percent. Typically, 3.5 cc of gas per A/h per cell per day of which 2.5 cc are hydrogen.

Comparison of Various Types of Cells

A simple comparison of battery types used for a UPS system is detailed below.

Prices are rated against Plante at 100 percent (which should give a 20-year life). All VRLA cells are rated at 10-year design life, the exception being the nickel-cadmium sealed cell, which should give a 20-year life. Rating is 26.6 kW, autonomy period 15 min, float voltage is 409 V, end of discharge is 306 V, or 180 lead acid cells float 2.27 VPC to end of discharge 1.7 VPC. Ambient temperature 20°C.

Cell type	Total cell price	Number of cells or blocks	Dimension (mm) L	D	H	Total weight kg	Total volume, m3
Plante	100%	180 cells each	190	133	212	2061	0.96
VRLA (AGM type)	23.8	30 × 12 V each	350	166	174	690	0.30
VRLA (AGM pure lead)	29.0	30 × 12 V each	197	169	173	477	0.17
VRLA (gel type)	27.4	30 × 12 V each	234	169	190	660	0.23
Nickel-cadmium, sintered plate	319.0	282 cells each	86	86	276	761	0.58
Nickel-cadmium, sealed	348.0	141 × 3.6 V each	145	195	406	2608	1.62

The above figures are at best a rough guide, and economics may vary with the power rating being considered

One should also bear in mind the life-cycle of the various cells. VRLA cells are normally available with a design life varying from 5 to 10 years, and there is the possibility of increasing these figures. At least one manufacturer claims a design life of 20 years, but under a strict regime of operation. In the writer's knowledge most 10-year design life cells afford in practice a life of approximately 7 to 8 years.

There is little doubt that an installation with a long forecast life (say 20 years) may find nickel-cadmium or Plante cells more attractive, but site conditions such as available space will have a significant impact on final choice.

One should also bear in mind that cost of battery replacement is high and attention should be given to labor costs, the disposal costs of the old cells, and downtime for the system, as well as to cost of new cells.

It should be remembered that all the figures and diagrams relate to manufacturers' accredited information. There will be slight variations between manufacturers' figures.

Future Trends

Future trends will show an increasing emphasis on relating battery design to UPS application.

A problem facing battery manufacturers is the ever-increasing desire to reduce the use of such elements as cadmium and lead, both of which are considered environmentally unfriendly. However, well-established and effective recycling mechanisms exist for all batteries using these elements.

Possibly, future UPS units may use lithium-ion cells which continue to be developed. Certainly, changes in design of batteries will result from the very active research program in progress at present.

Bibliography

IEEE 1184-1994—Guide and sizing of batteries for UPS
IEC 62060 Battery Monitors
UL 1778—Flame retardancy of battery cases
UL 94V-0 or 94V-2—Flame retardancy
UL 924—Battery vents
BS 6290 EN-28601
BS 6290 EN60068-2-6
BS 6290 EN 60068-2-20
BS 6290 EN 60068-2-32
BS - EN 60623

Books

Berndt, D., *Maintenance-Free Batteries: Lead-Acid, Nickel/Cadmium, Nickel/Metal Hydride: A Handbook of Battery Technology*, New York, Wiley, 1997.

Crompton, T.R., *Battery Reference Book*, 2nd ed. Butterworth-Heinemann, 1995.
Linden, D., and Reddy, T.B. (eds.), *Handbook of Batteries*, 2nd ed., New York, McGraw-Hill, 1999.

A great deal of technical support in writing this chapter has been gained from the following:

Thom Ruhlmann, C & D Technology, Milwaukee, Wisconsin

Colin Chapman, Manchester, and Kalyan Jana, Enersys, Warrensburg

Peter Hollingworth, Yuasa, United Kingdom

Chapter

8

Kinetic Energy as an Alternative Power Source

Introduction

Kinetic energy from flywheels has been in use for a variety of stabilizers and UPS units in the past.

Stabilizers with a short ride-through period consisting of a motor flywheel and generator may provide an isolated power source with a short ride through, the ride through being in the order of 1 s. The system consists of three shafts—motor, flywheel, and generator. It is not practicable economically to increase the ride-through time since the loss of the incoming supply and dependence on the flywheel inertia will result in a slowing down of the rotational mass and, thus, in the output frequency being out of tolerance.

Circuit Developments

A further use of the flywheel occurs with the original design of rotary UPS systems as illustrated in Fig. 8.1. Under normal operation the mains supply drives the dc motor via the rectifier. On loss of mains the system derives power from the battery. However, at the point of loss of mains the dc machine sees an immediate drop in dc voltage corresponding to approximately 15 percent. (Note this voltage drop will vary slightly depending on the type of battery employed and the autonomy period.)

The dc machine regulator is comparatively slow in response. Thus, the flywheel enables the system to tolerate the dynamic change, resulting in the ac machine output remaining in tolerance.

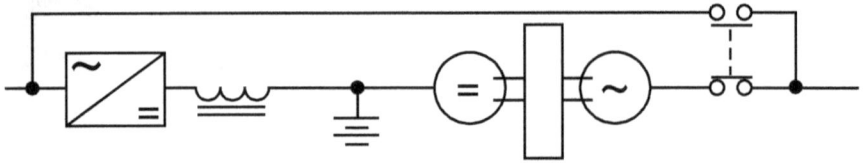

Figure 8.1

Over time developments have enabled us to reconsider the use of kinetic energy. Recent developments in circuitry and components have now made it feasible to consider kinetic energy as an alternative to chemical battery systems.

Clearly, any such alternative system must show advantages and also ensure the utmost reliability. Manufacturers must illustrate advantages in space, which is nearly always at a premium, and in its ability to form a reliable energy source for the stated autonomy time required by the circuit—usually the time between the loss of mains and the availability of full power from the standby diesel generator set. In addition, for the system to be economically viable, it must be efficient; there are losses both in the bearings and windage losses which normally are not inconsiderable.

Since the stored energy in a flywheel is proportional to the square of its rotational speed, manufacturers must require the flywheel to operate at much higher speeds than previous designs, and in this area advances made for other applications may be used for this field.

Quite a few new materials have been developed which will allow higher speeds with reliability and a list is shown below.

Material	Specific strength (in^3)
Graphite/epoxy	3,509,000
Boron/epoxy	2,740,000
Titanium and its alloys	1,043,000
Wrought stainless steel	982,000
Wrought stainless steels	931,000
7000 series aluminum alloys	892,000

Specific strength is a measure of material strength to weight density.

However, the costs involved in using many of these materials are prohibitive and in designs currently available steel is used, although this limitation may be solved in the future.

Variable-speed drives are now available which enable us to drive machines economically at much higher speed. So, circuits are available which can harness the kinetic energy inherent in high-speed flywheels and a typical circuit is shown in Fig. 8.2. Systems of this type are available over a large range of ratings from 30 to 1100 kVA.

Flywheel Type En Vacuo

One particular manufacturer has increased the system efficiency by operating the flywheel in a near vacuum condition and using magnetic levitation. The end result affords an efficiency comparable to the losses of a typical battery system under float charge. The construction is ingenious in that all wound components are at the periphery of the vacuum enclosure thus avoiding the build up of hot spots in the design. See Figs. 8.3 and 8.4.

The design incorporates the motor/generator/flywheel into one single piece of forged steel. The field coil provides the current to magnetize the teeth of the steel rotor which rotates past the the copper coils imbedded in the armature to generate power. Under failed power input the rotor slows down and the field is increased to raise the magnetism of the rotor teeth, thereby compensating for the speed loss. This in turn ensures a constant voltage output until 80 percent of the available rotor energy is utilized. The magnetic field flow is illustrated in Fig. 8.5.

The circuitry external to the flywheel housing is illustrated in Fig. 8.6. In this design the operating pressure is 50 mm tor and is kept at this pressure by an external vacuum pump which in practice is operational as occasion requires, probably called to operate on a 3 to 4 day basis. Losses are 1.75 kW for a system giving power to operate with a

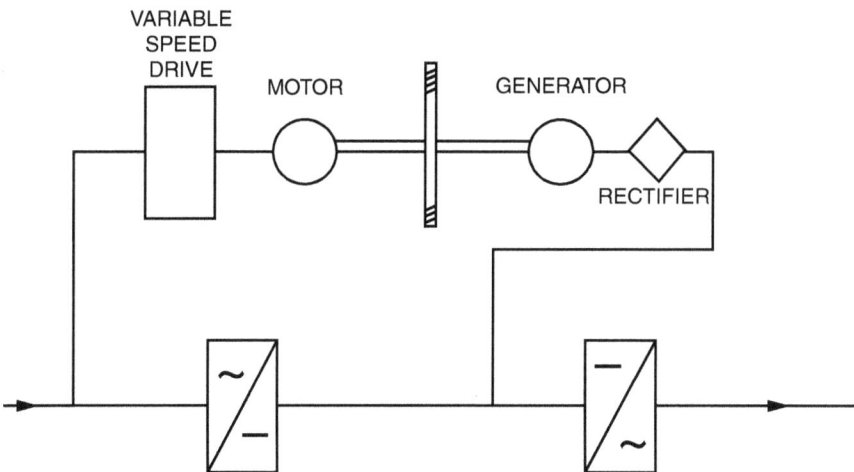

Figure 8.2

Cross Section View

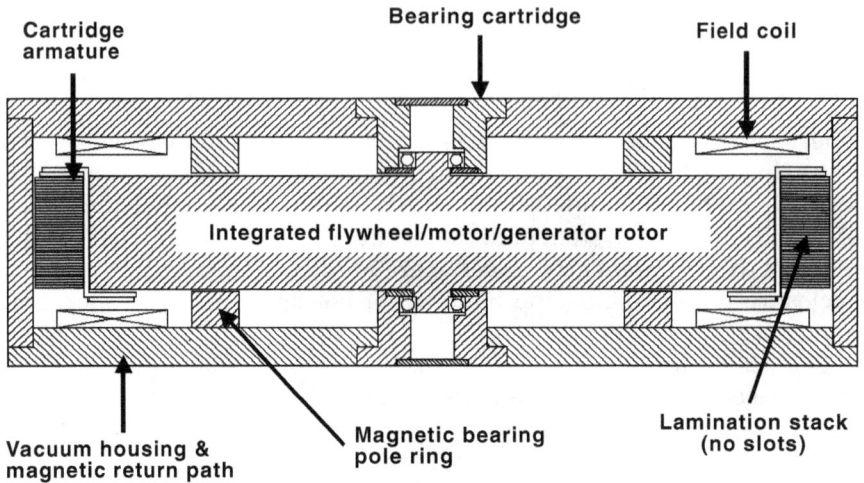

Cartridge armature

Bearing cartridge

Field coil

Integrated flywheel/motor/generator rotor

Vacuum housing & magnetic return path

Magnetic bearing pole ring

Lamination stack (no slots)

Figure 8.3   Cross-section view.

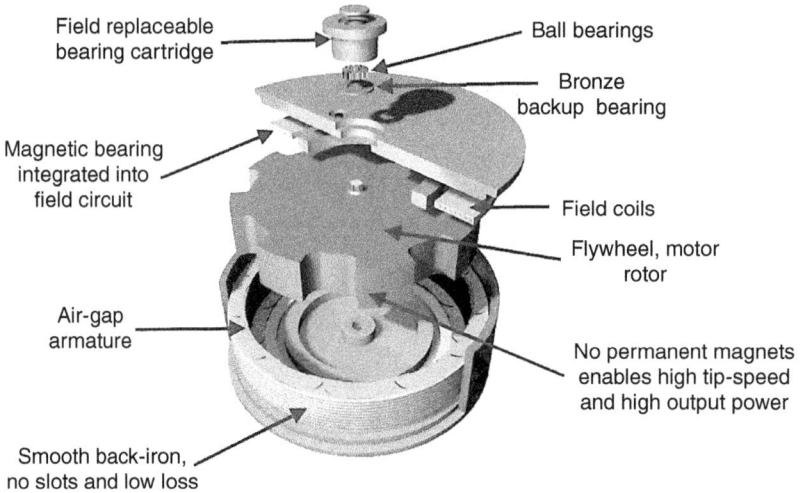

Field replaceable bearing cartridge

Ball bearings

Bronze backup bearing

Magnetic bearing integrated into field circuit

Field coils

Flywheel, motor rotor

Air-gap armature

No permanent magnets enables high tip-speed and high output power

Smooth back-iron, no slots and low loss

Figure 8.4

250 kVA UPS for 12 s. This is sufficient time to initiate a modern diesel generator set and obtain full power. Normal rpm is 7000, reducing at end of discharge to 2000. Such units are compact as seen in Fig. 8.7. This is a 250-kVA UPS system with the flywheel in the base of the unit. The overall size of the system complete is 1066.8 mm (42.0 in) wide,

Magnetic Field Flow

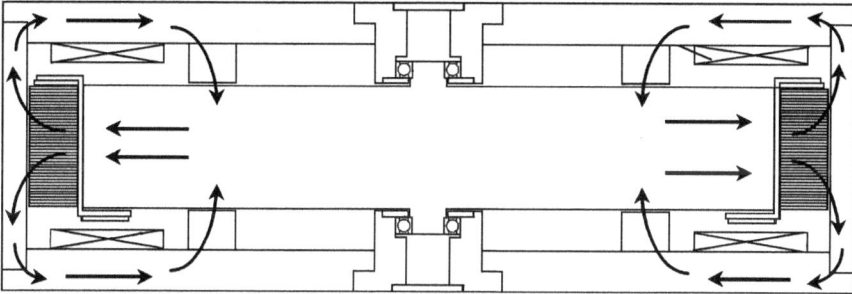

Figure 8.5 Magnetic field flow.

871.2 mm (34.3 in) deep, and 1976.1 mm (77.8 in) high, with a weight of 1700 kg (3750 lb). Operational noise level is 68 dBA at 1 m. Units may be connected to any static UPS module across the dc rail, and paralleling of the modules is available if required.

Flywheel Type In Helium

Another design is a vertical machine. In this case the rotating mass is enclosed in a sealed area flooded with helium gas. This clearly reduces friction losses and enables a good transmission of heat energy from the rotating mass. Again, magnetic bearings are in use during normal operation. Overall efficiency of the unit is 95 percent, and the normal working speed is 3600 rpm and at the end of discharge it is 1800 rpm.

Systems of this type are available from 150 to 1670 kVA and, clearly, autonomy periods vary from 2 min at 150 kVA to 12 s for a 1670-kVA load.

In this case, the unit dimensions are 2120 mm (83.46 in) wide, 1320 mm (51.97 in) deep, and 1900 mm (74.8 in) high, with a weight of 6000 kg (13,235 lb). Operational noise level is 74 dBA at 1 m. See Fig. 8.8.

So far as can be ascertained, the Piller design is in use with rotary systems but not yet with static systems. However, there seems to be no engineering reason to inhibit its use.

Claims for this system include a smaller footprint, savings in overall weight, lower maintenance, and the elimination of the use of hazardous materials, compared to an equivalent battery system.

Either system will operate over a wide temperature range—at least 0 to 40°C. Clearly, both designs above will be expanded in the future with other ratings. At this point installation costs are slightly higher

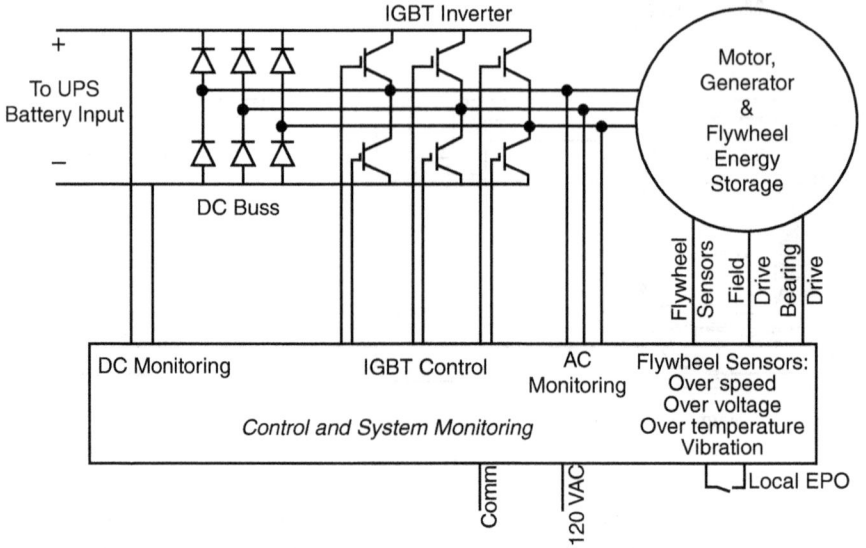

Figure 8.6

Figure 8.7

than for a battery complete with a full battery monitoring system. The advantage of reduced space is, however, attractive. As regards reliability, exhaustive testing has been carried out on the designs and at least 200 sets were in operation at time of going to print.

Acknowledgment

Technical support for writing this chapter was provided by Active Power Inc., Austin, Texas, and Anton Piller, Osterode, Germany.

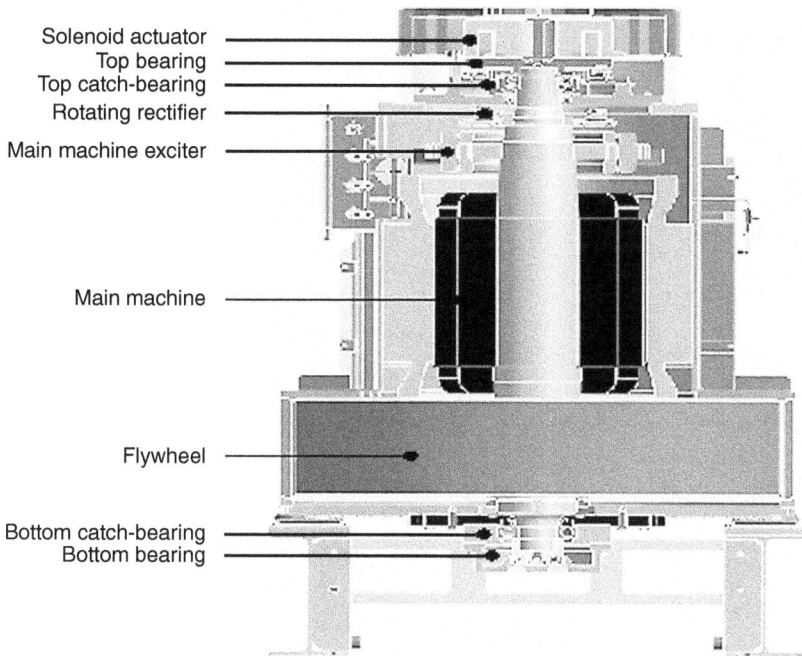

Solenoid actuator
Top bearing
Top catch-bearing
Rotating rectifier
Main machine exciter

Main machine

Flywheel

Bottom catch-bearing
Bottom bearing

Figure 8.8

9

Notes on Systems Installation

This chapter provides guidance on creating a complete system. Although not exhaustive it will give guidance and illustrate areas where problems may be envisaged.

The quintessential question is system power rating. It is surprising how hindsight illustrates how easily errors occur in this area. For example, the usage of computer systems inevitably grows with time, and often it is difficult for the exact rating to be determined. Here are a few points which need examination: The existing computer has every chance of expansion, but by how much and over what time period. Is it feasible that the computer may be reduced in size and much of its work is delegated to peripheral computer centers?

The possibility of adding further power blocks at a later stage, and thus paralleling UPS units and prime movers will have to be considered, it should be noted that if this decision is preferred the eventual system cost will be increased.

Although we have considered computers, the same considerations should also apply to other loads. Lighting has its own particular aspects requiring attention. Discharge lamps do have fairly high starting currents, and some discharge lamp circuits will necessitate consideration of waveform effect on the UPS and, indeed, the generator. It should also be remembered that even the mundane Tungsten lamp has a starting surge of approximately 12 to 15 times normal run current, and halogen lamps approximately 20 times normal run current. These starting surges are transient, lasting no longer than approximately 5 to 10 Hz, but it is advisable to take such surges into consideration.

Motor loads will usually give starting surges and here we are looking at surges up to 15 times normal full-load current and certainly the decay time to normal operation will be rated in seconds.

Consideration should also be taken of the well-known fact that predicted loads have a habit of increasing, when it is evident that a high security power supply is envisaged for a stated location.

Clearly consideration should be given as to the type of UPS system to be employed, static or rotary and which circuit is most suitable. As a guide, rotary sets these days are of higher reliability but at a significantly higher cost. Static systems certainly are almost universally in use up to about 50 kVA, and gradually beyond this rating rotary sets should be given consideration. Price and reliability have to be balanced for the project in hand.

Assuming that the UPS power rating has been ascertained, attention may now be given to the rating of the generator (usually a diesel generator set). Thus, consideration should be given to the following:

Computer rating and possible waveform problems

Air conditioning and variations in load due to operational requirements

Emergency lighting: there may be two sections, one portion directly from the UPS and (with a delay in operation) from the generator

Security systems

There may be other loads to consider depending on site conditions.

The generator will usually assume loads in the following way: initially air conditioning, lighting, and security. Note that it may be advisable to switch these loads on with small delays using a delay timer system to avoid heavy surge loads onto the prime mover. In most instances the UPS is the last load to be presented to the prime mover as it will be working from the battery system. The UPS also has a load ramp-up feature which allows a relatively slow rise in power demand from the UPS.

It should be remembered that the power demand from the UPS will consist of three sections: UPS load, losses in the UPS system, and recharging the battery. This extra load, present when recharging the battery, may be deferred until the system is restored to mains operation, but this deferment of battery recharge does have the penalty that a second outage of mains power soon after mains restoration may result in a complete loss of UPS power due to a battery incapable of supplying sufficient power.

Care should be taken to ensure that any harmonics induced onto the prime mover are taken into consideration. As stated in previous chapters harmonics induced onto the mains or the prime mover from the UPS will vary with UPS design. If, for example, a fifth harmonic filter is employed on the input to the UPS, low load conditions and variations in frequency from the prime mover may have a detrimental effect.

Under certain adverse conditions, such filters will exacerbate the waveform problem.

The static UPS system may be classed as a comparatively high impedance power source and thus adequate sub-subject protection should be allowed for, usually in the form of high speed fuses or circuit breakers. The many designs of rotary systems may have up to 14 to 15 times FLC to allow for subcircuit protection, so the problem is somewhat eased.

The autonomy period of the battery will have to be considered. As a guide most systems have 10-, 15-, or 30-min autonomy but there are some notable exceptions. Oil/chemical production, power generation, communications, and defense systems may have very long autonomy periods even up to 12 to 24 h, but these are the exceptions. The choice of 10 to 15 min autonomy has been made from the following considerations: It was considered feasible to shut down the computer in an orderly fashion if the prime mover failed to provide power. Since it will take at least 20 to 30 s for the detection of prime mover failure, is there sufficient time to react to such a condition and then perform an orderly shut down?

The alternative of employing wet cells should be envisaged and the following points should be taken into account. Stands, usually of steel construction these days, should be provided with a heavy plastic sheathing or a high-quality paint to inhibit chemical attack. The battery isolator should be immediately adjacent to the battery stand as cable from the isolator to the actual battery is clearly unprotected. Battery terminals are to be insulated usually with a close-fitting plastic cap. If cells are to be operated at relatively high ambient temperatures say above 30°C then the specific gravity of the electrolyte in the cells should be reduced (refer to the manufacturer).

Clearly, such cells will evolve gas when charging, in particular hydrogen, and a 4 percent admixture of hydrogen with air is an explosive condition. Most advice is to ensure that the maximum hydrogen content in the environment is no greater than 1 percent. The production of hydrogen is basically a function of plate area, and charging current.

A simple formula to calculate gas emission is as follows:

Number of cells $\times$ charging current $\times$ 0.0004194 = cubic meters.

Knowing the overall volume of the room, it is then easy to calculate the number of air changes per hour to ensure a low concentration of hydrogen gas. Such ventilation should preferably be by fans employing flameproof motors. The room should use flooring capable of withstanding acids, walls should be painted in a corrosion-proof light-colored paint, and lighting fixtures should be vapor proof.

You should note that VRLA cells do not completely inhibit the evolution of gas, it is simply significantly reduced and most ventilation systems will deal with this. However, emanation of hydrogen may be a problem for large VRLA systems and one should allow for roughly a tenth of hydrogen gas evolution compared to a normal flooded cell.

In the case of nickel-cadmium flooded cells, ambient temperatures for operation are between -20 to $+40°C$ (-4 to $+104°F$), with some increasing derating below $-15°C$ ($59°F$). Similar environmental conditions apply. Instead of protection against acid corrosion, the protection should be against alkaline corrosion. Again, the emission of hydrogen is calculated by using the above guideline formula. Note that in practice the gas emission from an alkaline cell is higher—for a given dc potential voltage across a system you need more cells than for lead acid. (Note: The nominal voltage for lead acid is 2 V per cell compared to nickel-cadmium 1.2 V per cell.)

The alternative to batteries is kinetic energy from flywheels. Such systems may be considered on grounds of space saving and possibly on the grounds of high ambient temperature. Such systems are relatively high in initial cost and insufficient time has elapsed for the writer to determine long-term running costs. The cost of bearings replacement against the known high cost of removing and replacing a battery. Power efficiency comparisons show no very marked differences, possibly a slight advantage in lower power consumption for the battery system.

Rotary UPS systems have their own advantages and disadvantages. If we look at the rotary transformer system then clearly we have considerations for the battery as defined above. The rotary transformer system has the advantage that the standard system has a low harmonic distortion to the mains supply (3 percent) well within G5/4. There are no starting surges, and harmonics induced onto the machine from the load can reach very high levels before any effect occurs.

Dimensionally the system is little different from a static system, but is heavier. The unit may be broken down into discreet sections for installation. Antivibration mounts are used and the units are not fixed to the floor. Maintenance is required (lubrication every 6 months), apart from this a yearly visit is advisable. As with static systems, remote monitoring is available.

The generator will clearly have to cover all load requirements and we would suggest that the following points are given consideration:

Discharge of exhaust gases

The warm and polluted air discharged from the engine room

Noise break out from the engine room and from the exhaust discharge

If there is a remote radiator, warm air will be discharged from it and it will be an additional source of noise

Vibration should be considered if the generating set is within an occupied building, steel framed buildings are particularly susceptible to noise generated by rotating sets

Leakage of fuel, fuel tanks are frequently required to be bunded

The generating set should be close to its electrical load and as close as practicable to a supply of fuel

An expansion of generator installation guidelines can be obtained in Chap. 1.

The chances of a prime mover failure are remote these days. For example, in lieu of batteries we now have a flywheel design which requires the prime mover to provide full power in approximately 12 to 15 s. Or indeed the close coupled diesel sets are able to support the system within 2 to 3 s!

The project may require consideration of local government authority, as regards the effect of appearance, noise, exhaust fumes, and of course local fire regulations. Consideration will also have to be given to mains power.

Adequate space for the system will require consideration to allow servicing of the equipment. It should be noted that in the case of most cubicles (either rotary or static UPS) rear access for servicing is required. Access to wound components such as transformers is often given a low access priority on the supposition that failure rate of such devices is low.

Since the devices are heat generating, adequate ventilation must be provided. For example, a static UPS rating 200 kVA will be a fan cooled cubicle generating some 16 kW of heat at full load. The UPS itself can work in ambient temperatures up to 40°C. However, the supporting battery, particularly those of a VRLA design, is best operated at 20 to 25°C (68 to 77°F). When designing battery stands or enclosures ensure that cooling air is allowed to surround the cell by allowing a gap of 10 mm between cells. Heat build up in such enclosures can be a major defect. Typically, an enclosure will allow a 2 degree ambient temperature rise.

The design of small UPS modules where space is at a minimum can have a major effect on battery performance. Also, ensure that for such small units operational staff refrain from locating the device adjacent to a heat source such as a radiator, and that the usual warning lamp of low battery is visible!

It is surprising how many instances we have found of the small UPS being installed in unfavorable conditions, and then forgotten! There is a tendency for such units to be treated as another black box, just ignored and then blamed when called upon to supply power during an emergency and failing. The problems here are as follows:

A tendency for operational staff to ignore them as a passive object seemingly not contributing anything to the task they are faced with

Such units are often hidden from view and the tell-tale warning lamps are either ignored or not even seen!

Small units suffer from a lack of maintenance and indeed it has to be understood that maintenance costs for such systems may be exceptionally high, for example, replacement of a battery may be uneconomical.

Clearly, solutions to these problems are fairly straightforward. Advise operators of the qualities of such units, place units in a position where they can monitored, and regularly check them. It should be noted that many such sets may be interrogated remotely electronically.

An unusual application for UPS systems is railway signaling. Clearly, such loads are vitally important. Any loss of power on busy routes may cause not only dangerous conditions but it may take days to reorganize traffic flow. The availability of a suitable power source trackside is not always immediately to hand. Some areas requiring power can be remote and so the answer has been to use the raw power available to the electric traction units. Such power sources are very often 25 or 33 kV with some wide variations in regulation: ±30 percent has to be considered and surges have been reported up to 60 kV. In some instances, recordings have indicated a twice crossover of 0 volts in one-half cycle. Clearly a power supply fraught with problems!

To work under such conditions, the UPS has to be protected against wide voltage variations and surges, this is achieved by well-designed isolating transformers and surge diverters. The wave trace shown in Fig. 9.1 illustrates input, and the output voltage (upper trace) which is approximately 650 V single phase.

Designs vary depending to a great extent on variations to be expected in the input voltage range. One school of thought is to use standard UPS units well protected against the vagaries of the high-voltage supply and illustrated in a photo of a system using two Silcon/APCC 40-kVA modules with isolating transformers and controls (see Fig. 9.2). The other approach is to use a specially designed UPS unit again with full isolating transformers and controls. Where such wide variations in supply are encountered a normal bypass circuit is clearly unfeasible, and one alternative is to use a a hot standby UPS (i.e., a UPS in working mode but off load) switchover via a static switch. Figure 9.3 is of a specially designed 60-kVA UPS to operate from a 25-kV power source and with IP54 construction.

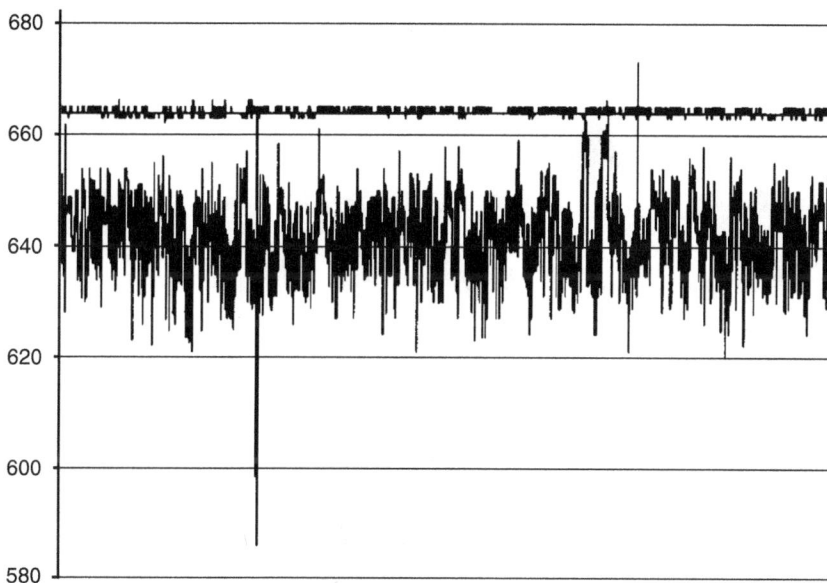

Figure 9.1   Wave trace.

Where wide input voltage variations are to be expected, a normal by-pass system may not be used. In such cases the bypass is obtained from a standby UPS system. If this is the chosen design, then the standby system must be in an operative mode that is inverter functioning.

Other problems to be expected with such systems are the need to remember that most such systems are usually adjacent to the rail track and may suffer from EMC emitted by the raw power from the overhead lines or from passing traction units.

Many such installations clearly have to contend with wide tempera-ture variations and other problems such as dust. Dust may contain iron from brake shoes. Air-conditioned steel-clad containers are often used to house such equipment. The steel cladding assists in providing a fara-day cage to inhibit EMC problems.

Whereas the main market is endeavoring to reduce the utilization of wound components, an area where the use of full isolation from the incoming supply is still essential is in the medical field. UPS units are often provided to guarantee power to respirators, heart pumps, and other medical apparatus that is connected to the patient and any power problems may affect the sensitive medical monitoring systems. The UPS under such conditions should comply with EN 60601 or UL 2601; this implies a 4000 V test and also that earth leakage is nil. To comply with this the solution usually adopted is to use a low-impedance

Figure 9.2   Universal power.

transformer whose secondary has one side tied to a good earthing system (see Fig. 9.4).

Small, geographically isolated telecommunication repeater stations very often require a high reliability of power although no mains supply is available. Thus, such stations rely on either wind or solar energy with, of course, an energy store (battery) to ensure a continuity of power to the load (see Fig. 9.5).

The geographic location of the system will determine whether wind or solar energy is in use. In either case weather will have a large effect on the availability of power and the use of a battery is usually inevitable. The choice of the type of battery is subject to many constraints. In most instances, the battery will perform under quite large variations in temperature, and deep cycling of the cells may be expected. Neither characteristic is available from the VRLA battery. In practice we have found that either nickel-cadmium or the tubular lead

Figure 9.3

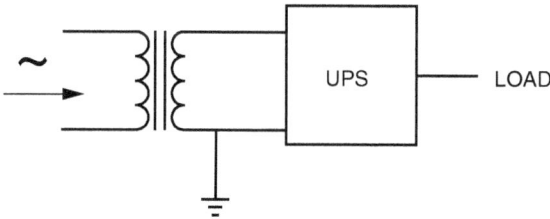

Figure 9.4

acid cell is preferable. For reasons of initial cost the tubular cells should be considered. Such cells are available with a large volume of electrolyte, enabling the cells to withstand long periods without electrolyte replenishment.

The dc-ac inverter illustrated in Fig. 9.5 may act as either an inverter feeding power to the load from either wind or solar or battery source. In the event that these sources are unavailable, the generator supplies the load and the inverter reverses its operational mode, acting as a charger for the battery system. Note that reverting to the generator is unusual, but there is no loss of power during the changeover period.

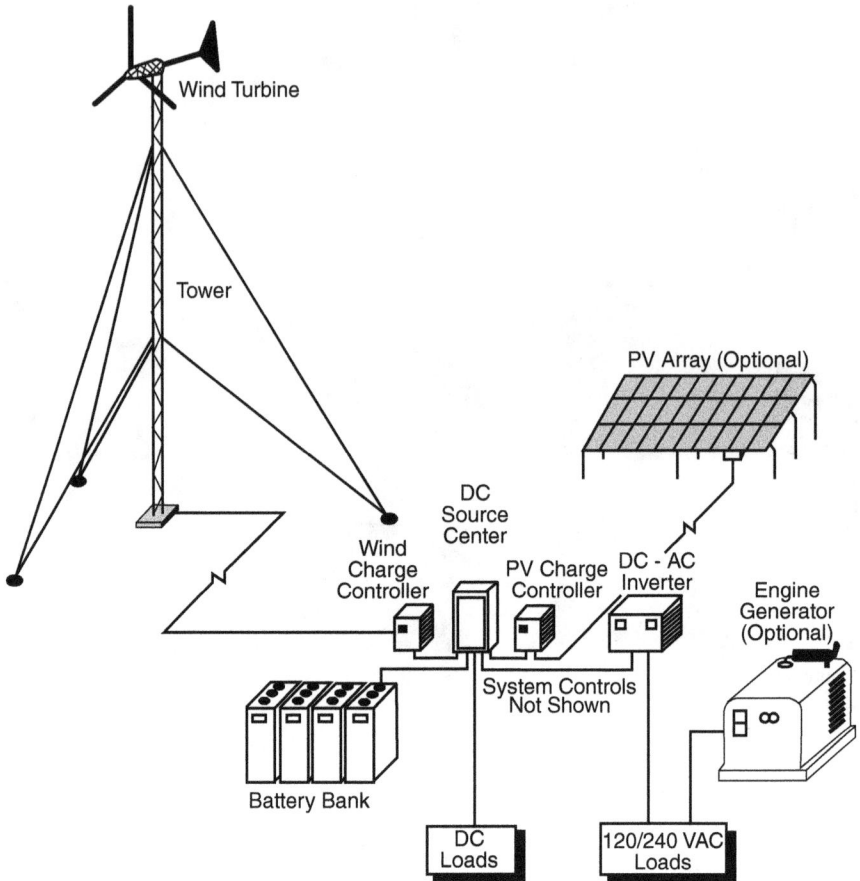

Figure 9.5

Local advice should be obtained regarding both the availability and reliability of wind and solar energy and a cautious attitude taken as regards to periods of absence of either source!

Airport runway lighting is clearly a critical load and the most advanced specification calls for no outage to be longer than 1 to 2 s (Category 3). However, in many cases UPS units are in use.

There are instances where the use of standby emergency lighting systems is precluded due to the very nature of the lamp load. We refer to high-pressure discharge lamps, the inner envelope of which, when quiescent, has a very low gas pressure and it is relatively easy to strike a discharge through the low gas pressure tube. This gas pressure builds up when the lamp is running and full light output is attained at possibly 1 or 2 atmospheres pressure. To restart such a lamp on power failure is almost impossible. The lamp has to cool down (possibly taking 3 to 5 min) and then the lamp has to be restuck and, clearly, full lumen output will take up to 10 min to achieve. Thus, it may be preferable to utilize a full UPS system.

The choice of static or rotary UPS systems is, in the writers' experience, difficult. Higher rating systems will tend to favor the rotary systems. The choice will depend on many considerations: reliability, cost, maintenance, dimensions, site conditions, and, in some cases, even access to site will have a bearing on the final choice.

Table 9.1 gives an indication of actual size of systems and estimated efficiency. It is based on an imaginary system—the load consisting of a 50/50 split of power demands: 50 percent essential and 50 percent noncritical load. Three ratings were chosen: 400, 1000, and 2000 kVA. The dimensions included are for the generating sets and control, but no fuel store, and in the case of the static UPS alternative a 10-min battery and a 12-pulse input rectifier are included. In the case of the rotary transformer system, a kinetic energy system was utilized. The dimensions are comparative based on the static UPS at 100 percent.

The flywheel design for the rotary transformer system is overrated for the ratings chosen and, thus, distorts the figures to a certain extent. It should also be remembered that initial price and running cost will favor the static alternative.

Maintenance for larger sets does need attention. Nowadays, many sites are constantly monitored by the manufacturer's own service department. Alternatives are available for regular maintenance checks usually every 6 to 12 months by qualified staff. It is important to thoroughly investigate that such contractors are supported and trained by the equipment manufacturer. We recommend that a thorough investigation of servicing arrangements be made prior to equipment purchase.

Table 9.1

	Floor dimensions	Weight	Overall system efficiency
Static UPS 400 kVA	100%	100%	91%
Rotary transformer 200 kVA	165%	156%	94%
Generator/clutch/m/c 200 kVA	178%	75%	95%
Static UPS 1000 kVA	100%	100%	91%
Rotary transformer 500 kVA	165%	64%	94%
Generator/clutch/m/c 500 kVA	99%	56%	96%
Static UPS 2000 kVA	100%	100%	91%
Rotary transformer 1000 kVA	112%	64%	94%
Generator/clutch/m/c 1000 kVA	80%	48%	96%

Acknowledgment

Supporting technical information for this chapter was given by Universal Power, Loughborough, United Kingdom; Powernetics, Loughborough, United Kingdom; and Bergey Windpower Co., Norman, Oklahoma.

Some System Failures: The Light of Experience!

Introduction

It is part of an engineer's education to experience an occasional failure. The incidents which are briefly described in the following paragraphs have been included in this book in the hope that readers will benefit from them. They demonstrate the need for lateral thinking at the planning or design stage and the truth of the dictum that if it can happen it probably will.

Lack of Ventilation

This failure occurred at a prestigious multiset installation having a rated output of several megawatts several years after it had been commissioned. Test runs had been conducted at regular intervals and the operating personnel were confident that it would perform satisfactorily when it was required to do so.

A prolonged supply failure was, however, to prove their confidence misplaced; the sets started and supplied the load but after about 20 min there was a complete shutdown due to overtemperature. The reason was surprisingly simple, the duct carrying the engine-room ventilation air had been blanked off and there was no air flow. It is believed that contractors working in the winter had blanked off the duct for the benefit of their workers and had departed without removing it.

This failure demonstrates the need for test runs to be on load and of sufficient duration for thermal stability to be reached; until this failure, test runs had been of short duration.

A Bypassed Radiator

This failure occurred at a conventional single-set installation having a rated output of about 1 MW. The installation was one of many similar but not identical installations at various locations spread over the United Kingdom.

On loss of the normal supply the set started and supplied power but after a short time tripped out on coolant overtemperature. As in the preceding example, the reason was simple but was less easily explained. The coolant pipework had been repaired recently and had been installed incorrectly; the coolant flow bypassed the radiator!

This failure probably occurred during a test run but it demonstrates the need to conduct a test run after any major work has been completed.

Lack of Fuel

This failure is so simple that it barely seems worth recording but it is a real-life situation, an example of what actually happens! An important set was regularly tested on load, but when it was required to start following a mains failure it went through its multiple cranking sequence and registered Fail to Start.

The daily service tank was not automatically topped up and test runs in the past had drained it. There was no other fault. The daily service tank should include clear, visible indication of its contents, and its contents should be checked after each period of running.

Changeover of Supplies without a Break

There was a large number of identical small sets installed at locations spread over the United Kingdom. Some sites experienced frequent tripping at the start of test runs whereas others had no such problems.

It was found that at sites which experienced failures the test procedure involved starting the standby set, opening the normal supply switch, and immediately closing the standby supply switch. At the sites which had successful test runs the first two operations were conducted in reverse order so that the sequence was to open the normal supply switch, start the standby set, and close the standby supply switch.

At the sites which had successful test runs the starting of the standby set ensured a delay between opening the normal supply and closing the standby supply. The load at all of these sites included an ac electrical rotating machine and the failures demonstrated the need for a short delay (a few seconds) to allow magnetic fluxes to decay when changing from one supply to another unless the supplies are synchronized. Some machines will take a very heavy transient current if they are connected to a supply while running at or near synchronous speed, in such cases the delay should be long enough to allow deceleration to, say, half speed.

Restoration of Supply to an Inertially Loaded Drive

This event concerns a large (probably 200-kW) squirrel cage induction motor coupled to a centrifugal fan of welded sheet steel construction. The large inertia of the fan resulted in a long run-on time after a supply failure, and it was found that restoration of supply, even after a long interval, resulted in overload tripping.

This incident illustrates a different principle from that illustrated by the previous example. An induction motor may be considered as a transformer, the stator winding being the primary and the rotor winding the secondary. If a supply is restored to such a machine whilst it is running at, say, synchronous speed, there is a sudden increase of stator rotating flux which is stationary with respect to the rotor. The rotor winding will be seen as a short circuit and the stator flux is diverted into the leakage flux paths, resulting in a large stator current which causes the overload trip. If tripping does not occur, the flux transfers to the rotor iron, and the stator current reduces, at a rate determined by the rotor time constant.

The above description is a simplification of the effect, the motor is likely to be running at a subsynchronous speed which will result in a low-frequency current in the rotor, but unless the reader is already familiar with the phenomenon it is easier to consider the behavior when synchronous speed applies.

Large motors with large inertial loads need special attention if they are likely to be energized at speeds above, say, one-half of the running speed.

Low Transformer Oil Level Due to Low Ambient Temperature

This failure occurred at a conventional single-set installation located in an exposed location well above sea level. The generator was connected to a generator transformer which was located outdoors without weather protection.

During a particularly cold spell of weather, the normal supply failed and the standby set started, but it immediately shut down due to operation of the Buchholz relay which indicated low oil level.

The Buchholz relay had operated correctly, at normal ambient temperature the oil level would not have been so low, but the reduction of oil volume due to the low temperature had caused the oil level to fall below the Buchholz float chamber. Most outdoor transformers are continuously energized and do not experience a cold ambient temperature but standby generator transformers are idle for most of their life and are only occasionally energized.

This failure indicates the need for the designer to be aware of the unexpected! This incident was followed by a change to the specification for generator transformers to ensure that a visible and adequate oil level is maintained down to $-10°C$.

Inadequate Protection against Driving Rain

This incident was an extreme inconvenience and a potential failure. The site was on the west coast of Scotland and the generator had a rating somewhat above 1 MW.

The ventilation air inlet had the usual weatherproof louver which faced seaward and toward the prevailing wind. In normal circumstances the louver may have been adequate, but at this location the generator room was the on the ground floor of a two- or three-story building. Above the inlet louver there was a high blank wall which caught all the driving rain, which then ran down the wall toward the louver. The louver had not been designed to cope with the large quantity of water and much of it was drawn into the generator room.

The building should have incorporated some feature which diverted the rainwater from above before it reached the intake louver, a form of guttering for instance. In addition, it would have been expedient to build a wall or other construction to prevent driving rain from reaching the louver. This failure demonstrates the need to be aware, at the planning stage, of potential problems which may not be obvious. This installation would undoubtedly have passed its usual commissioning tests, but it would not have been raining heavily at the time!

Unconventional Use of a Harmonic Filter

The installation provided power for an office block serving a financial institution. It comprised a static, uninterruptible power supply rated at 120 kVA and a much larger standby set rated at 1 MW. The load was the usual mix of communication, computing, and display equipment.

The uninterruptible power supply was unusual in that it included a harmonic filter connected to the UPS output, the purpose being to reduce the harmonic content of the load to comply with Engineering Recommendation G.5/3 when using the bypass supply. A contactor enabled the use of the filter to be controlled. The UPS connections were as indicated in Fig. 10.1.

The filter comprised, for each phase, three series resonant circuits tuned for the third, fifth, and seventh harmonics, and connected across

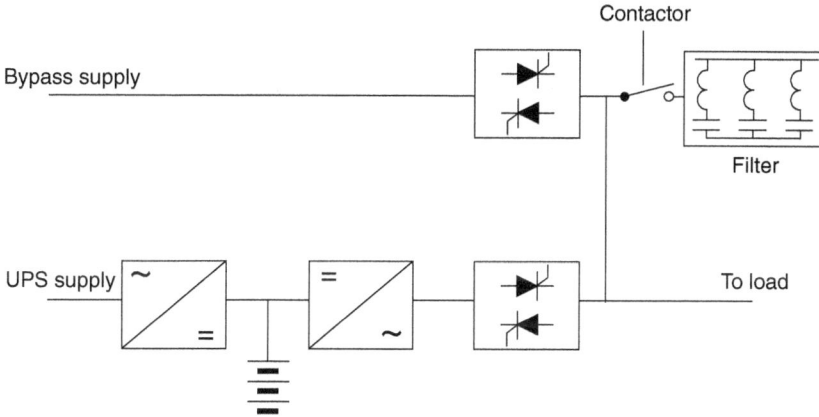

Figure 10.1   Block diagram relating to unconventional use of a filter.

the supply. Such a filter is seen by the supply at fundamental frequency as a capacitive load and the component values were such that with a sinusoidal supply a fundamental current of 167 amperes would flow, equivalent to a rating of 120 kVA.

On two occasions, separated by several months, the user's equipment connected to the supply suffered damage due to overvoltages; on the first occasion the damage was extensive and disastrous, on the second occasion, less so. The origin of the overvoltages was not immediately apparent but was found to be due to incompatibility between the filter's capacitive current and the static switch. The circumstances of the two occasions were entirely different but both occurred at weekends when the electrical load connected to the installation would have been low, and both were associated with the use of the bypass supply.

When the building load is normal, the connection of the filter has only a small effect and the installation sees a load current with a fairly high power factor, but when the building load is small and the filter is connected, the installation sees a load current with a leading power factor close to zero. The gates of the two thyristors forming the static switch are opened alternately for alternate half cycles; if a leading current flow is demanded before the appropriate gate is opened, there is no path available for the current. One thyristor is in reverse current mode and the other has its gate closed, severe distortion is inevitable. In fact overvoltages of 1000 V were measured—three times the normal peak voltage of the sine wave supply.

This episode is included as it indicates the need, at the planning stage, for perception and for the need to avoid unconventional arrangements. It demonstrates the advantages to be gained from taking advice from the manufacturers who have learned these lessons from experience.

An Unstable Power Supply

This incident occurred at a remote location supplied from an overhead line connected to a diesel engine generating station. The equipment of interest was a rotary-type of UPS in which the synchronous machine was supplied with power from a line commutated inverter. On two occasions a fault on the overhead line was followed by a commutation failure within the equipment. It was important to find an explanation for the commutation failures and a reconstruction of the sequence of events follows.

The overhead line failures would have been phase-to-phase or phase-to-earth faults probably caused by bird strikes. The supply voltage would have been severely depressed and at that time several dynamic features would have come into play:

- Energy would have been drawn from the UPS flywheel for a few tenths of a second until the battery contactor had closed, the rotor speed is therefore reduced leading to a low output frequency.

- When the battery contactor closed, the inverter would immediately go into current limit in order to accelerate the rotor and correct the low frequency.

- The generating station voltage regulator would have increased excitation in an attempt to restore the voltage.

As a result of these dynamic features, when the fault was finally cleared the supply voltage from the overhead line suddenly increased to 118 percent of nominal and this voltage was applied to the line commutated inverter already in current limit. Commutation was not possible and the failure led to a short circuit of the dc supply and a complete failure of the UPS output.

The equipment included a rectifier contactor, and to prevent a recurrence the control circuit was changed to ensure that on a loss of supply the contactor opened and remained open for 2 s. The equipment was quite capable of accepting the 118 percent voltage surge under normal conditions and the short delay ensured that on restoration of supply the inverter would have come out of its current-limited mode of operation.

This incident demonstrates the complexity of the occasional failures experienced on systems. It is not always easy to reconstruct the events leading to a catastrophic failure. After this incident the fault conditions were reproduced within the manufacturer's works to demonstrate that the reconstruction was credible.

An Overenthusiastic Charging Regime

This event occurred many years ago when flooded cells were in use. The conditions are unlikely to be repeated nowadays, but a description may serve a useful purpose.

The installation included a conventional UPS using a nickel-cadmium battery as its energy store. The purchasers of the equipment had, against strong advice from the manufacturer, insisted on an unrealistic charging regime which required a boost charge after every supply failure. It is for this reason that nickel-cadmium cells were used instead of the usual lead acid cells.

One weekend (when the installation was not attended) the supply authority was working on the local distribution system and there was a series of interruptions of short duration. As a consequence, the already fully charged battery was subjected to a continuous boost charging rate.

After many hours there was an impressive failure, one or more cells exploded and devastated the battery room. The electrolyte level may have been low at the commencement of the boost charging. There was no way of knowing, but it is believed that the electrolyte level dropped during the charging due to the predictable electrolysis. As a result of the electrolysis, the cells became full of an explosive mixture of hydrogen and oxygen and the atmosphere within the battery room was probably overloaded with a hydrogen/oxygen mixture.

As the electrolyte level continued to fall, there came a time when the electrolyte was level with the bottom of the plates; the inevitable spark occurred causing the explosion.

The situation would not have arisen if valve-regulated recombination cells had been used because boost charging of such cells is not allowed. The incident is described here because it indicates a hazard of which readers should be aware; it also indicates the danger of failing to consider the probable effect of using an unconventional charging regime.

Loose Intercell Connections On a UPS Battery

A series of winter storms resulted in an irregular mains supply to a large computer network in a city center. As the disturbances were erratic and of short duration, the supporting generating set was not started, thus over a period of a few hours the battery was called on to support the load whenever the mains was absent.

This resulted in a serious fire enveloping the entire battery room. Site personnel were made aware of the situation by the UPS in a separate room; in fact the UPS registered wide dc voltage variations which the system interpreted as a battery failure. Sufficient warning was given to ensure that the fire did not spread from the battery room. Needless to say, the entire computer network was shut down.

The resulting investigation established that the three parallel strings of 400-V 400-Ah flooded lead acid cells had not been properly maintained. Maintenance logs only recorded that the cells had merely had electrolyte levels topped up. There was no record of torque testing the cell terminals. It was found that the terminals on some cells were only finger tight! The heating and cooling caused by the discharge cycles had caused further deterioration of the already bad connections and eventually arcing had occurred, initiating a serious fire.

This incident indicates the importance of proper battery maintenance. Before passing on to the next incident, spare a thought for the individual who had to enter that battery room to disconnect the system!

An Unsuccessful Attempt at Cost Reduction

A project engineer had to design a system comprising a static UPS, a generating set, and the usual ancillary equipment to complete the installation. After the system had been installed it was found that the generating set was unable to support the load in bypass mode.

In unraveling the problem it was discovered that the project engineer had approached several suppliers for a complete installation and had chosen a particular supplier, only to be told by his management that the cost was too high. As a result he decided to design the system himself, but with insufficient knowledge of the component parts. The load included a large number of office-type PCs with a high harmonic content and a high crest factor, measured at the generator as 3:1. The generator and its voltage regulator had been chosen without reference to the harmonic loading.

This failure indicates the hazards of not employing a main contractor with responsibility for coordinating all aspects of the installation. An experienced contractor, before accepting responsibility, would undoubtedly have asked questions about the load and its harmonic content.

Empty Sumps

Exporting generating sets to developing countries can lead to interesting problems! Two sets were shipped for a computer center and installed by local staff. On commissioning, both sets failed due to there being no lubricating oil in the sumps. Generating sets are shipped with drained sumps, and the need to add oil had been overlooked.

Lack of Cooling Air

A large city bank installed a multimodule parallel redundant UPS system which was commissioned without problems but failed within a week. The load included a large number of office PCs and, as the failure was caused by overheating of a large output transformer, it was suspected that the harmonic currents drawn by the PCs had caused overheating. Measurements of the load current taken adjacent to the transformer indicated a 3:1 crest factor which is within the capabilities of most static UPS.

Further investigation revealed that the overheating had nothing to do with harmonic currents, it was due to a lack of cooling air. The UPS units had cool air fed to them from under the floor, exhaust air being emitted from the top of the cubicles. At a central T junction the underfloor duct was joined by the main cool air duct, and the transformer was mounted above the T junction. It was intended that the cool air would flow left and right to the cubicles and upwards to the transformer, but it was found that very little air flowed upwards hence the overheating of the transformer. Modifying the T junction improved the air flow and resulted in satisfactory operation of the UPS.

It is not easy to check every detail at the commissioning stage, but the air flow rates should really have been checked at some time before putting the equipment into service.

An Inadequate Supporting Structure

These two incidents concern the close-coupled diesel/clutch/kinetic store/generator type of UPS, both being installed within steel-framed buildings and supported on steel joists.

At one site problems were experienced during installation due to the deflection and curvature of the supporting joists. To overcome the problem it was decided to install steel columnar supports beneath the set. Fortunately there was a firm base able to take the load, but the columns were inconvenient, restrictive, and expensive.

At the other site the set was installed without any known problems but there followed a series of bearing failures which were attributed to the resilience of the mounting.

The size and weight of these machines makes major repair or maintenance very difficult and structural problems are best avoided. These events occurred long ago and lessons have been learned. However, if such machines are to be supported on other than a solid foundation, it would be wise to discuss the installation with the manufacturer.

Index

absorbent glass mat (AGM), batteries, 193
accelerating transient condition, generator motor, 94
accessibility of generating set, 45, 113–114
Accu rotor rotary UPSs, 172, 174
acoustic treatments vs. noise, 111–112
acoustic louvers, 112, **112**
active power filters, harmonic distortion, reducing, 136
air conditioning loads, 232
air cooled diesel engines, 26–27
  ventilation systems and, heat loss and, 29, **32**
air flow vs. heat loss calculations, ventilation systems, 28–29
airport runway lighting and UPS
alarms and warning devices, 41–42
aldehydes, as exhaust gas pollutant, 38
alternating current generators, 2, 9–13
  connections in, delta/star or star/interstar, 12–13
  excitation systems for, 10–12, **11**
  kilowatt rating of, 10
  pole face damper windings in, 12
  single– vs. two–bearing types, 10
  speed (rpm) of, 10
  standard reference conditions for, 13
  steady state short circuit current in, 78
  transformers in, 12–13
  voltage regulators for, 13–14, **14**
alternative energy, 239, **240**
aspiration, diesel engines, air requirements, 5
automatic starting and control systems, 40
autonomy period of battery, 233
average power calculation, diesel engines, 3–4

backpressures, exhaust systems, 34
balanced current systems, 80
balanced load, static UPSs, 153
barometric pressure
  diesel engines and, 7
  gas turbines and, 9
  batteries, 19, 185–222, 235, 238–239
  absorbent glass mat (AGM) in, 193
  ampere hour rating for, 144
  autonomy period of, 233
  cell life vs. number of discharge/charge cycles, 200–201, **202**, 235, 249

barometric pressure (*Cont.*):
  cell types, 185
  charging systems/cycles for, NiCad, 214–220, **216, 217, 218**, 214
  charging systems/cycles for, lead acid, 195–205, **197, 201, 202, 204**, 235, 249
  chemical reactions and basic design, in lead acid, 186–187
  comparison of various types of cells in, 220–221, **220**
  connection point for charger in, 205, **207**
  constant voltage charging in, 195–205, **197, 201, 202, 204**
  containers or cases for, 205, **206**
  corrosion in, 204–205
  cost of, 220–221, **220**
  electrochemical efficiency in, 200, **202**
  electrolytes in, 191, 193, 195, 209, 217, 220
  environmental considerations for, 233–234
  exploded view of, lead acid, **189**
  future trends in, 221
  gassing in, 203–204, **204**
  gel type VRLA cells in, 205–206
  intermittent charging in, 196–205, **197, 201, 202, 204**
  lead acid, 185–191
  loose intercell connections in, 249–250
  loss between charge/discharge, in lead acid, 186–187, **186**
  materials used in plates, 204–206
  monitoring and trend graphs, 208–209, **210, 211**
  nickel–cadmium (NiCad), 209–220
  parts of, **194**
  pasted plate cells in, 187–191, **188, 191**
  Plante cells in, 187–191, **188**, 193
  plate in, **190, 195**
  pocket plates in NiCad, 212, **212**
  rechargeable, 185
  rectification in, 198–199
  ripple in, during charging, 198, 200, **201**, 215
  rotary UPSs and, 169
  sealed type NiCad, 214, 217, **219**
  sintered plates in NiCad, 213–214, **213**
  starved electrolyte state in, 193
  static UPSs and, 144, 163, 208

Note: **Boldface** numbers indicate illustrations.

253

ABOUT THE AUTHORS

Alexander King and **William Knight** are private consultants specializing in UPS and standby supplies. They reside in Gloucestershire and Surrey, United Kingdom, respectively.

www.ingramcontent.com/pod-product-compliance
Lightning Source LLC
Chambersburg PA
CBHW060349220326
41598CB00023B/2859